Preston Watson
1880–1915

Preston Watson
1880–1915

Dundee's Pioneer
AVIATOR

Alastair W. Blair
Alistair Smith

Librario

Published by

Librario Publishing Ltd.

ISBN: 978-1-909238-48-0

Copies can be ordered via the Internet
www.librario.com

or from:

Brough House, Milton Brodie, Kinloss
Moray IV36 2UA
Tel/Fax No 00 44 (0)1343 850 178

Printed and bound in the UK

Cover Design: Monika Gromek

Typeset by 3btype.com

Preston Watson 1881–1915

Dundee's Aviation Pioneer who beat the Wright Brothers into the air by at least five months in 1903.

Many believe that the Dundee born Preston Watson beat the famed Wright Brothers into the air by a margin of several months in the early years of the 21st Century.

Eye witnesses interviewed in the 1950s, clearly recalled the flight setting off from a primitive landing strip at Errol by the banks of the Tay in the summer of 1903.

Watson's wire and wood flying machine was hoisted by means of ropes and weights into the trees, catapulted with engines running, and flew some 100–140 yards before landing.

Encouraged by his success, Watson went on to build two further planes.

He joined the Royal Flying Corps at the outbreak of the first World War but was killed in a training accident at Eastbourne at the age of only thirty-four. He was buried on July 5th, 1915 in Dundee's Western Cemetry.

Preston's brother James campaigned for his brother's feat to be acknowledged, but the authorities were sceptical.

Now, however, Dundee's first pilot is finally receiving the recognition he deserves.

THE INSCRIPTION ON THE PLAQUE
IN THE RECEPTION AREA OF DUNDEE AIRPORT

Contents

Foreword

The object of writing this book is not to mount claims about who first flew with powered, dirigible, heavier-than-air machines; although, like some others, the authors consider it likely that Watson was "off the ground" around 1903/4. The object is rather to put together as much information as possible; some unpublished; some previously published in books, magazines and newspapers, especially in his native area – and to collect this in a single publication. Much of the material is already known in diverse quarters and the authors found when researching down various avenues, that many of the paths had been trodden before although some had become somewhat overgrown again. A good deal of it is inevitably anecdotal and an attempt has been made to construct the text, and to supply references and acknowledgements in a manner which indicates the quality of the evidence.

As will be evident, the text involves the comparison of differing opinions and this has necessitated some repetition for the sake of clarity, but which some readers may find somewhat irritating.

It is not our intention to discredit opinions held by others, except in the very few cases where hard evidence exists; but rather to challenge what may have hardened, with the lapse of time, from opinion to perceived fact; and to offer in some cases alternative interpretations of what ultimately goes back to the same basic material.

It may seem that some challenge is offered to some established authorities, notably the late aviation historian Charles Gibbs-Smith and this may seem an impertinence. He is not now able to re-engage in further argument but the authors feel that, as a professional and an academic, he would have been glad to engage in the cut and thrust of renewed discussion – without rancour on either side.

It is hoped that the book will, at the same time, document and celebrate the life and achievements of one who should be considered a key figure in early aviation. It seems probable that had he not been killed at a relatively early age, his aviation career might have projected him into a position with the more celebrated early pioneer aviators but, like Percy Pilcher before him, a promising life was claimed by the very activity that offered the promise.

Alastair W. Blair
Alistair Smith
March 2014

CHAPTER I

Introduction

The development of all branches of science and technology depends on an accumulation of observations and information which tends to grow exponentially, leading to a level of understanding sufficient to make solidly-based advances and is seldom the work of a single individual – except perhaps at the very beginning; and that which is hailed as a 'first' is usually the apex of a mountain of preceding endeavours by others.

So it is with the development of heavier-than-air aviation. Since time immemorial man must have envied the birds their ability to fly and realised that flying was a possibility for heavier-than-air creatures. (In the whole timescale of history, of course, it is only recently that man realized that air is more than the absence of substance). It is no surprise that the earliest attempts by man to fly were made by imitating the birds. But this began in an era when scientific method was undeveloped and those early imitations were not based on an understanding of *how* birds flew.

With the advent of the industrial revolution there was an increasing understanding of mechanics and fluid mechanics, and an academic approach to the development of scientific method combined with the realisation of the importance of theoretical principles. This brought about an understanding of how a wing derives lift from its motion through the air. Certainly things began to happen in aviation and it is possible to trace back some of the very roots of that science. What is odd is that there seems to have been little or no connection made with what must have been a fund of practical experience with sails going back to pre-history, and keels going back to at least the 11th century.

However, the general ethos of the time was towards the existence of 'inventors' and high academic (even mad!) professors making their inventions in isolation in the search for personal renown. This probably fostered an atmosphere of secrecy among rivals and impeded the creation of a growing pool of experience and sound evidence-based knowledge. In aviation, of course, the outcomes of such experiments were often rather public!

There have now emerged a considerable number of designs and actual attempts to fly by individuals in the 19th century. Some were based on a rudimentary understanding of the physics involved and some resulted in fledgling flight. Some were doomed to failure (often spectacular) and some merely remained as plans which, by modern analysis, might have been capable of flight.

Those listed in *The Story of Flight* [1] can be grouped accordingly:

Those credited with some sort of flight:

Felix du Temple	hot air engine monoplane	1874
Alexander Mozhaisky	steam monoplane	1884
Clement Ader	steam monoplane	1890
S. Pierpoint Langley	steam model tandem quadruplane	1896

Planes not constructed		
W. S. Henson	Ariel steam carriage	1842

There are others, many not well reported and in the first few years of the 20th century there was an increasing number with varying degrees of success.

The foundations for manned flight can be traced back further. The French were the main experimenters with balloon ascent going back to the Montgolfier Brothers (although the hot air balloon was probably a Portuguese invention). The first recorded manned balloon flight took off from Paris in 1783. (The first manned aerial flight of any sort in Scotland was a balloon ascent from Edinburgh by Tytler in 1784 and the first balloon free-flight was by Vincento Lunardi across the Firth of Forth from Edinburgh to Fife in 1785). The first convincing *dirigible* flight was that of Henri Giffard in an airship, powered by steam, which flew in 1852.

The principles involved in gliding were worked out by Sir George Cayley (1773–1855) [2] who devoted his life to the study of aviation as it then was, and his extensive work laid sound foundations for those who were to follow. It was the German, Otto Lillienthal, who took matters forward to the brink of powered flight. His long experience in the development of gliders, with painstaking records and calculations, certainly played a seminal role in the development of gliders and were a major influence on Percy Pilcher and others in Europe and the Brothers Wright and others in America.

The Wright Brothers' experiments with aerodynamics in wind tunnels were to a great extent *ab initio* and Samuel Franklin Cody came to aircraft construction from his passion for flying kites and his efforts in making man-lifting kites; nevertheless, it should be remembered that experiments on lift and drag had already been undertaken by Cayley in 1804! Crucially, the Wright Brothers' wind tunnel experiments led to a re-appraisal of a good many of Lillienthal's calculations.

Up to this time, attempts at powered flight had used steam or hot air engines for propulsion. The internal combustion engine was already in existence by later in the 19th century although not yet able to provide a sufficient power/weight ratio for flight.

The question of who first achieved powered flight was a matter of great individual and national importance at the time, and some of the rival claiming led to heated disputes. There seems little doubt that it had been achieved in some imperfect way before December 1903, but as is often the case in such matters, authentication is difficult or impossible. In the case of flying, what constituted a 'flight' had yet to be defined. Although Jullien flew a clockwork driven model airship in 1850 and Langley flew his steam driven model in 1896, the interest centred on who was first to achieve manned flight, with power take-off (perhaps assisted) in a heavier-than-air machine whose direction and height could be controlled and which ended in a landing (or at least the possibility of one). Then there was the problem of what distance across and above the ground qualified for the term 'flight'.

One could define 'flight' as a powered and manned aircraft lifting off from the ground and sustaining that flight through the air by means of the propulsion, before landing the aircraft in a controlled manner. The distance and height achieved are probably academic, but the important criterion is the sustaining of the flight.

A large enough body of collected experience did not exist at that stage and no independent organisation had the sapiential authority or the credibility to arbitrate any conflicting claims, so anything of that nature would have had to be done retrospectively. The Wright brothers had by far the best documented and witnessed claim to a flight which was piloted, took off under its own power and flew convincingly clear of the ground (for 852 feet in 59 seconds) on 17 December 1903. It was piloted by Wilbur who demonstrated that he had control of lateral stability and climb/descent; and was able to land without too much damage to pilot or machine. Without formal process this was generally accepted as the standard by which other claims would be judged. The speeds, altitudes and effectiveness of control all moved on from then. The Wrights themselves did not consider they had satisfactorily conquered these aspects until 1905 – the year their patent was granted. Even then, they did not demonstrate their 'Flyer' in public until 1908.

Subsequent claims to have flown prior to the Wright brothers have been accompanied by less convincing documentation and history has come to accept the Wright Brothers as being the first to achieve the criteria for what now might be regarded as 'flight' in the aviation sense.

In one sense, it does not matter greatly whether one considers flight was first achieved by du Temple in 1874, Ader in 1890, Mozhaisky in 1884, Whitehead in 1901, Watson in 1903 and many others, or the Wright Brothers in 1903. Without wishing to diminish in any way the determination, genius and bravery of any of these (undoubtedly best documented in the case of the Wright Brothers), it is true to say that they were able to do so working from a foundation built up by others. That view is reinforced by the fact that these men, working largely independently, achieved their aim in some measure all within the space of twenty-eight years. Cayley, Lillienthal, Pilcher and others had brought matters to the cusp of success. But the efforts of the empiricists such as du Temple and Mozhaisky, and even some whose efforts with tie-on or mechanical flapping wings engender mirth and ridicule, made a contribution.

Compiling a comprehensive review of the activities leading up to the achievement of manned heavier-than-air flight in the late 19th and early 20th centuries is well beyond the scope of this publication, but it is clear that there was such activity in many parts of the world, and some of the major texts such as those of Charles Gibbs-Smith [3] and a number of other well researched and referenced texts are listed in the bibliography (which makes no pretence of being comprehensive). It may help to give a flavour of the scene to consider briefly *some* of those activities and to underline some of the difficulties of making retrospective historical assessment.

The object here is to attempt to set the achievements of Preston Watson into that context in the hope of giving him the place that the authors and many others feel he deserves in the story of pioneer aviation.

CHAPTER II

The Background To The Times

The popular perception is that the accolade for the first flight goes to the Wright Brothers, Wilbur and Orville, using a heavier-than-air flying machine of their own construction and an internal combustion engine of their own design and flown by Orville at Kill Devil Hills, Kitty Hawk, North Carolina, USA on 12 December 1903.

As with many of the pioneers, the matter became something of an obsession with them, and they closely observed soaring birds, studied what was already written on the subject and conducted their own experiments with aircraft and apparatus of their own devising. Wilbur in particular pursued the matter with single-minded determination. Given that neither brother had been through higher education or had any training in research methodology, it is doubly impressive just how thoroughly and ingeniously they tackled the problems; how meticulous they were in recording their results; and how sensible they were in observing safety measures.

Sons of a bishop in the Brethren Church, they had had a kindly but austere upbringing and tended to eschew publicity and grandiosity and avoided exaggerated claims of success. The construction of a functioning flying machine was their objective – not impressing the outside world with their achievements or making themselves a large fortune. In the end they achieved powered flight, attracted world acclaim and accrued modest wealth.

It was Wilbur who first developed an interest in aviation and at first the brothers pursued their interest, not in secrecy, but without attracting much notice. At the end of the 19th century, flying kites and the sight of two young lads attaching aerofoil shapes to their bicycle handlebars would not seem so remarkable.

In 1899, Wilbur wrote to the Smithsonian Institute – a foundation committed to promoting the accessibility of scientific knowledge – and asked for relevant information on heavier-than-air flight. At that time communications in general were profoundly slower, less efficient and less complete than they are now but it is likely that by this route the Brothers learned of Lillienthal and others. There is a touch of irony in the fact that the Secretary of the Institute was Samuel Pierpoint Langley who was actively engaged in constructing his own flying machine at that time. It is difficult to know how much was known by one experimenter about any of the others unless they actually published (as Lillienthal did). However, a rather unusual enthusiast – a Chicago engineer named Octave Chanute – maintained a keen interest in developments and promoted contacts between many of those working in the field. The name of Octave Chanute is well known amongst aviation historians but as far as more general acclaim is concerned, he is an undersung hero. His greatest contribution to aviation was not so much the application of his undoubted engineering ability, but his activities in collating the efforts – failures and successes – of those actively engaged in the field and putting those involved in touch with each other as far as possible. He sponsored some experimentation in gliders and built some to his own designs. He wrote advocating attention to safety considerations. But most importantly, he acted as an academic journal in human form.

In 1894 he published a book [4] entitled *Progress in Flying Machine* which could be used as a benchmark. So much so that it was last reprinted in 1976, although by that time it was of historic significance only!

He corresponded a good deal with Samuel Pierpoint Langley of the Smithsonian Institute.

History may criticise Langley for not distributing knowledge on the subject as well as he might have done. It was after all, the stated purpose of the institution of which he was secretary. He did, however, respond to Wilbur Wright's approach of 1899 and through him the Wright Brothers were in close touch with Chanute. As close, that is, as they would allow the contact to be. Alas, subsequently some suspicion arose that Chanute was claiming for himself some of the credit for their progress in the field, and relations became strained.

Chanute was born in France, moving to America with his parents as a young child. The original name was Chanut – but when pronounced the French way, it translated to 'naked cat'. Octave added the 'e' to the spelling to obviate this. He became an experienced and sought-after engineer mainly building railways and bridges. He used some of his own wealth to fund his own glider experiments and to sponsor the work of others.

Being a French speaker and well known in aviation, he was invited to address the Aero Club de France in 1902. On that occasion he described some of the progress being made by the Wright Brothers and warned the French that if they wished to be first in the field, it would behove them to hurry. As mentioned elsewhere the French seemed to ignore the warning; indeed many in France ignored the activities of the Wrights, hailing Alberto Santo-Dumont as the first in the world with his flight, near Paris, of 1906.

Samuel Pierpoint Langley assumed the post of Secretary of the Smithsonian Institute via an interest in astronomy but during his long term of office he became interested in flying machines. His proposed solution to the problem of stability in the air was the provision of sufficient power – under-emphasizing the matter of directional control, lateral control in particular. While he was pursuing (vicariously through Charles Manly) a power unit of favourable power/weight ratio, the Wright Brothers were examining the forces of lift and airflow in order to understand how to control a flying machine in flight.

Langley's steam-powered unmanned model made a credible flight of half a mile or so in May 1896. This convinced him (and he convinced his financial backers) that his approach was sound and he went on to construct a manned machine, large enough to carry a pilot. Unlike the Wrights, he did none of the hands-on construction work himself but left it to the members of the teams that he brought together. As well as problems with the airframe design and construction, the provision of a suitable power unit proved difficult. Manly spent a good deal of time on this aspect and his success in the matter was notable. Based on a design of Stefan Balzer's which had not fulfilled its promise, he refined and developed a power unit with a surprisingly good power/weight ratio which continued in use for some time afterwards.

At first Langley did not seem to take very much notice of what the Wrights were doing but, possibly through Chanute, he learned of the progress made and this may have spurred him on to complete the project and to stage a test flight. This took place amidst heavy – and sometimes hostile – publicity owing to Langley's irascibility in general and with press relations in particular.

Unfortunately, two consecutive attempts ended in wreck no further than the end of the launching rails. The second of these took place only days before the Wrights' successful flights at Kitty Hawk eclipsed Langley's efforts. Despite the belief on the part of Langley and of Charles Manly that these failures were due solely to a steel pin on the launching mechanism and that otherwise the aircraft would have flown, Langley made no further attempts. This may have been due wholly or in part to the fact that his funding was withdrawn.

Although the Wright Brothers were in the air within a few weeks of Langley's attempts, Langley's military backing was not transferred to them. It is likely that their low profile up to this point (compared with the august standing of the Smithsonian Institute) meant that no consideration was given to the matter in the relevant quarters. When the Brothers, seeing the military potential of a flying machine, subsequently tried to interest the American Government, the initial response was one of no interest. It may have been Langley's costly failure that crushed their interest – for the time being at least.

Tobin [5] in his book *First to Fly* describes these events in detail and the image conveyed of Langley is that of a man of high academic standing, busy with the work of Secretary who was vicariously conducting the flying machine project, but whose approach to the project was surprisingly unacademic and non research-based. He travelled a good deal and enjoyed the company of prominent scientific figures in other countries. He was aware of the French aspirations to be at the forefront of aviation development, but he did not appear to pay much heed to anything happening in Britain at this time.

He also was apparently unaware of the activities of Gustav Whitehead (aka Weisskopf). Whitehead's background was sailing ships and subsequently kite construction. He claimed to have built and flown a steam-powered monoplane and flown it near Pittsburg in 1899 for a distance of about half a mile at a height of twenty-five feet before failing to clear the side of a building – a claim corroborated by his assistant, Louis Darvarich, who suffered burns in the ensuing crash.

Whitehead continued, helped by modest financial backing, and flew another steam-powered monoplane in 1901. This flew for about half a mile, Whitehead steered to avoid some trees and made a landing by cutting the power. This flight took place in Connecticut and was witnessed *inter alia* by the (adult) son of the proprietor of a local newspaper, the *Bridport Herald*. The story was also carried by the *New York Herald* and *the Boston Transcript*. No attempt was made to claim a 'first' but at the time there was no obvious body to grant such an honour. Fuller details are published by the Historical Flight Research Committee Gustav Weisskopf [6] but Gibbs-Smith [7] roundly dismisses Whitehead as a fabricator and calls the existence of these flights into serious question.

Given that Langley and the Wright Brothers were both striving to build a flying machine, it does seem curious that the newspaper reports were not picked up and brought to Langley's attention.

Matters accelerated after the 1903 flights and failures. Alexander Graham Bell, inventor of the telephone and himself of Scottish origins, was a long-standing friend of Langley. After the latter's high profile lack of success, Bell felt that he could take matters further, by concentrating on the construction of a kite that would be stable in most wind conditions and could be made large enough to support a man and an engine. He assembled about him a team, with Bell family financial backing and succeeded in making some advances. However some of his younger colleagues (notable amongst them was Glen Curtiss) constructed a flying machine which, after some revision, was flying by 1906. Curtiss' interest was to some extent commercial and he later co-founded a company and made a reasonable fortune out of aircraft production.

The Wright Brothers sued Curtiss for infringement of patent and were eventually successful. The bone of contention was wing-warping, which Curtiss abandoned and devised the aileron system instead. He is sometimes credited with their invention, but it seems to have been Robert Esnault-Pelterie who first devised and used recognisable ailerons in 1902. The challenge was to devise a means of banking and turning an aircraft, as one would do on a bike, so that it was not skidding through the turn. The ailerons were mounted on the outer ends of the wings and tilted the aircraft over into the direction in which the turn was going to be made. Combined with the action of the rudder, this enabled a balanced turn to be carried out.

Meanwhile interest in aviation was reviving in France. Santos-Dumont shifted his attention from airships to heavier-than-air machines and flying his number 14bis at Bagatelle in the Bois de Boulogne. This was soon followed by Farman, Voisin and Bleriot.

British Interest

General interest in aerial matters in Britain began in the 18th century with the balloon ascents. Tytler's ascent in 1784 and Lunardi's crossing of the Forth Estuary from Heriot's Hospital grounds in Edinburgh in 1785 (43 miles) were respectively one and two years after the original one in Paris.

As in France, interest developed in heavier-than-air machines through kites and gliders. Among those engaged in such activities at the end of the 19th century, in the years leading up to the achievement of credible aviation, was a young marine engineer, Percy Sinclair Pilcher (1867–99). Pilcher possessed a keen mind and after serving some years at sea worked for Glasgow shipbuilders and was eventually appointed lecturer in the engineering department of the University of Glasgow.

He began building manned gliders in 1891 and using his accumulating experience, built several successive gliders. He had studied Lilienthal's reports and made contact with him, eventually spending some time with him and actually flying one of his gliders. Pilcher's fourth glider (he called it 'the Hawk') seems to have been the most successful. It is known that he was preparing an engine to fit it. Unfortunately, while demonstrating the Hawk as an unpowered glider in 1899, it suffered a structural failure and crashed, injuring Pilcher so badly that he died two days later. The 'Hawk' is now displayed in the National Museum of Flight in East Lothian.

One relatively unsung pioneer of that era was Horatio Phillips (1825–1926). Phillips was a remarkable inventor and was interested in VTOL (vertical take-off and landing) and in autogyros. He initially worked with steam power, but later constructed his own internal combustion engines. His major contribution was his experimental work with wing design. As early as 1884, as a result of his own experiments, he patented a wing design in which the curvature of the upper surface was greater than that of the lower, inducing greater lift. This remains the basis of wing form to the present time. His multiplane aircraft were not, however, a conspicuous success although his second one made a flight of 150 yards in 1907.

A large determining factor affecting support for, and interest in, aviation was the view taken of its military significance. Graf Ferdinand von Zeppelin's rigid dirigible powered airship flew in 1900 and was enthusiastically developed in Germany. The British military establishment was at first dismissive of any military use for flying machines, but increasingly began to realise the possibilities in reconnaissance work. During the Boer War (1899–1900) the British used towers for observation and signalling. Static balloons dispensed with the demands of time and materials to build the towers and offered a more moveable alternative with the advantage of greater heights. Eventually dirigibles were used with some success.

Recognition by British Military authorities of the potential of the balloon in warfare is evidenced by the establishment of the Balloon Factory at Chatham in the early 1880s. One of the young engineers involved in manufacturing the first British military balloons was John Edward Capper. Capper later served in the Royal Engineers in the Boer War where he was involved with the deployment of the dirigibles. He returned to England in 1902 with the rank of Lt. Colonel RE and was appointed to take charge of the Balloon Factory. His significance in aviation history stems from his breadth of vision in seeing the potential of simultaneous experimentation in balloons, man-lifting kites and heavier-than-air machines; and the advantages of combining the knowledge and skills involved.

He must have faced opposition from certain sections of the conservative military establishment. It is known that Haldane, the Secretary of State for War was contemptuous of the possibility of powered flight and any possible military importance. Despite this and his day-to-day responsibilities, Capper promoted the activities of John Dunne in building gliders and eventually powered aircraft.

John William Dunne (1875–1949) had been invalided back from South Africa. He was the son of a prominent military household. When he developed an interest in building gliders, Capper employed him in the Balloon Factory in 1906 as the army's official researcher into the possibility of powered heavier-than-air flight. There is no doubt that it was powered flight that Capper was after – he was converted to the concept after meeting the Wright Brothers in America in 1904. He befriended them on a personal basis and the two families corresponded regularly at a social level.

Dunne was transferred, for the sake of secrecy, to the Blair Atholl estates in Perthshire in 1907, and ghillies were employed to keep the press (and spies) at bay [8].

His first powered glider (designated D1) was launched from a trolley in 1907, piloted by none other than J.E. (now Colonel) Capper, RE, crashing into a wall about eight seconds after launching – slightly injuring the person, and perhaps more significantly, the dignity of Col. Capper.

D3 suffered a similar fate, this time without Capper aboard. Dunne's first powered machine had two Buchet engines and was considered underpowered. It performed better with a 25hp REP engine but only managed powered hops.

The Army continued to support the research until 1909 when finance was withdrawn, by which time Dunne had not achieved credible independent flight. He continued the work with the specially formed Blair Atholl Aeroplane Syndicate (1910). His D5 design was subsequently built by Short Brothers for the Syndicate and flew convincingly in 1911. Several other models were produced.

A D8 was purchased by the Nieuport Company and a demonstration was conducted in France in 1912 in which the pilot climbed out on the wing while the aircraft was in flight. The designs were developed under licence in North America resulting in military aircraft for the First World War, but Dunne himself retired through ill-health in 1914.

Samuel Franklin Cody was certainly not working in secrecy. He was a legendary character whose undoubtedly harsh early life had made for a romantic figure – fearless, handsome, competitive and swashbuckling. His background was in the Wild West shows which toured America in the late 19th century, Cody having spent his early adult years in the saddle driving cattle hundreds of miles to the meat markets. At that time his name was Samuel Chowdery. His biography *'Colonel' Cody and the Flying Cathedral* by Garry Jenkins [9] gives a fascinating account of a remarkable man.

Cody attracted the attention of Capper in 1902 because of his work with man-lifting kites and throughout his subsequent work with powered aircraft he seems to have favoured the box kite configuration devised by the Australian, Lawrence Hargrave. His first successful plane was very large – hence the nick-name 'Flying Cathedral'.

Capper watched developments with interest and Cody was encouraged with some financial support (apparently often difficult to extort) and eventually with an appointment to the Balloon Factory. There was an objection to his working there as a non-British national which was eventually circumvented by awarding him the honorary rank of colonel. 'Honorary' really meant that his pay and employment were not secure; and his national origins, coupled with his other means of earning a living (the stage), and his lack of basic military training resulted in a prejudice against him on the part of the more entrenched establishment.

Despite some antagonism, some aeronautical set-backs and his exposure to the public directly and via the press (he enjoyed impressive public popularity), Capper was able to see that progress was being made at least as rapidly as that of his other prodigy, Dunne.

On 16 October 1908, Cody flew his 'Flying Cathedral' for 1,390 feet at an estimated height of 30–40 feet and appeared to have a degree of lateral control. The flight ended abruptly when wind turbulence caused it to crash.

As previously mentioned, in 1909 the military establishment prevailed and although funding was continued for the development of kites and balloons, the financial backing for heavier-than-air flight was withdrawn.

Cody did not take the Oath of Allegiance to the Crown until 1909 and therefore his claim to be the first Briton to fly was disputed, particularly by A.V. Roe. The credit for the first flight over British territory by a British-born pilot was subsequently awarded to Moore-Brabazon for a flight in 1909.

Alliot Verdon Roe (1877–1958), having won a competition with a model aircraft, constructed a biplane in which he made several powered hops. He claimed to have made a flight of 150 yards with it in 1908 and that this constituted the first British flight, although he himself considered that his first viable aircraft came later in 1912. A.V. Roe continued with aircraft design and founded A.V. Roe & Co. Ltd. in 1910, which continued with aircraft production through both World Wars with the famous Avro range of aircraft.

In a Scottish context it is worth mentioning here the three brothers Barnwell from Balfron (near Stirling), sons of a shipyard owner in Glasgow. They visited the Wright Brothers and returned to start building their own version in 1907–08, outwith any military association. They were 'off-the-ground' by 1909 and had achieved sustained flight by 1911 [10, 11].

French Interest

France certainly has a distinguished place in the pioneering of aviation. Although not the originators of the hot air balloon, the Montgolfier brothers made demonstrations to a wide public in Paris in September 1783 when an ascent was made carrying several farm animals as passengers.

The Aero Club de France was very active, if perhaps a little divided between those who believed in aeroplanes and those who believed in airships. What bound them together, however, was a determination that if anyone was to achieve aviation, it should be a Frenchman. It was certainly a very patriotic group. When visited by Chanute in 1902 they were keen to know what progress was being made in America – but the tenor of the proceedings gave the impression that they wanted reassurance that France was not being overtaken.

Several of the influential members at that time were entrepreneurs in the promotion of manned, powered flight in France, e.g. Leon Delagrange and Earnest Archdeacon, both of whom had first-hand practical experience. There is no doubt that some members were worried by accounts of the progress made by the Wright Brothers – but others doggedly chose to ignore them. This was in part possible because (and this was typical of the Wrights) they had not theretofore been publicised in America.

Prize moneys were offered for various 'first' achievements and this may have helped to keep matters going because although there was much talk, French progress seemed to stall for a while; Ferber, actively encouraged by Archdeacon, being the only one conducting attempts. Matters did start to pick up and the Voisin Brothers, Gabriel and Charles began building aircraft commercially in 1904, starting with gliders and later machines for Delagrange and Henri Farman.

Farman flew an airship in 1904 but did not start with powered flight until 1907 when, using a Voisin-built aircraft, he won prize money put up by Archdeacon. He was probably the main design influence in the aircraft he commissioned from the Voisin factory. These aircraft gained an enviable reputation for sound construction and relative stability and safety. So much so that Farman aircraft were extensively used in the First World War as bombers and gun-carriers rather than simply for observation and reconn-aissance [12].

Louis Bleriot (1872–1936) started building model gliders in 1902. His early attempts at powered machines were unsuccessful and it was not until after 1907 that he achieved powered hops of up to 150 metres. Thereafter, of course, he went on to take his place in aviation history with his cross-channel flight in 1909, thus claiming the *Daily Mail* prize for the first airborne crossing.

But in the atmosphere of patriotic enthusiasm for France to lead the field, it was Alberto Santos-Dumont who held the limelight. He was a Brazilian who came to Paris in 1898. His early endeavours were with airships and he started immediately on arrival, completing two in that year. These were powered by a small petrol engine of 3.5 hp. In 1901 he was very much in the public eye when he flew one of his airships (fitted with a larger engine) from the suburbs of Paris, round the Eiffel Tower and back in order to claim the Deutsch prize which was funded by Henri Deutsch de la Meurthe.

Santos-Dumont became a very popular figure in Paris. He was small of stature and according to his biographer, Wykeham [13] he was not only a charming, bon viveur but thoughtful, intelligent, practical and diligent. Certainly his charm and (under-stated) achievement comes over in his own account *My Airships* [14].

In 1906, he tested a powered aircraft of his own design by suspending it from one of his airships. Later, he made a powered take-off with this ungainly 'canard' type machine (labelled No 14bis) but only managed a short flight. In October 1906, with a larger engine fitted he flew 197 feet, enabling him to claim prize money put up by Archdeacon. In November of the same year, he flew 722 feet to claim the prize money offered by the Aero Club de France for the first plane to fly more than 100 metres.

All this received great publicity in France and was hailed enthusiastically as the first manned, powered, sustained and controlled flight in a heavier-than-air machine, despite the fact that it was two years and eleven months after the Wright Brothers' flight at Kitty Hawk.

The engines used by Santos-Dumont are of some relevance. According to Wykeham [15], the earliest was a 3.5hp de Dion engine. Santos-Dumont and his associate Chapin, contrived to link two of these together in an unusual configuration to produce a more powerful unit. This was used in one form or another in the first four airships. In 1901 he was using a 15hp Daimler-Benz air-cooled engine.

He won the Deutsch prize in airship No6 using a 20hp water-cooled engine described as a Buchet/Santos-Dumont. In 1902–4 he had a 60hp water-cooled Clement engine in airship No7 which was used again on No10. For superstitious reasons there was no airship No8, but No9 was a smaller craft with a 3.5hp air-cooled Clement.

The next few in the series were either not engined or never flew. No14 was another small vessel with the 3.5hp air-cooled Clement. With No14bis, he had started into heavier-than-air machines and from then until No19 in 1907, he used various Antoinette engines when he changed to using Dutheil-Chalmers engines, air-cooled flat twins which he promptly modified to produce more power – as was his habit. He produced a modification which he had manufactured in the Darracq factory – a modification which increased its horsepower rating from a nominal 20hp to a nominal 35hp. This engine fitted to his No20 ('Demoiselle') was flying in 1909. Further modifications took the series to No22.

The story has a tragic ending. About 1910 Alberto Santo-Dumont was diagnosed as suffering from multiple sclerosis. He became progressively more reclusive and erratic. In 1914 he destroyed his own personal records, including those relating to the design and construction of the airships, and in 1932, took his own life. A detailed account of his later years, along with insights into his character, his daemons and his claims to be 'first to fly', is given in a later, well-researched biography by Hoffman [16].

It is difficult to know, more than a century later, how much the question of secrecy entered into the course of events at the time. It may have operated at different levels – military, financial or personal rivalry.

In some quarters personal rivalry and competitiveness ran high. The race was on to be the regarded as the first to fly. Despite Chanute's efforts some of the front runners were very reluctant to share experiences and report their progress. The nearer the Wright Brothers came to success, the more careful they were to shun publicity and prevent what we now term industrial espionage.

In Europe this aspect was perhaps less marked. Although the French were very anxious to lead the race, they published much of what was taking place in the journals of the time.

From the financial point of view, filing for patents by hopeful inventors was more or less routine. In America the Wrights were very conscious of this aspect and were quite secretive about the details of their technical achievements especially between 1903 and 1906. Orville, after Wilbur's death in 1912 did not do a great deal to advance aviation, but spent much of his time trying to interest governments in their aircraft and fighting litigation actions over perceived patent infringements. Many of these were against Glen Curtiss [17].

On the military side, the Wrights had seen military potential early on and were bitterly disappointed in the difficulties they encountered in interesting the US Government. They had no scruples about approaching foreign governments, but even there they only met with a limited response.

The military considerations in Europe were different. France was aware of the increasing industrial and economic strength of neighbouring Germany; and British relations with the Kaiser were beginning to deteriorate. British military invincibility was impugned by the Boer Wars. Much of this stemmed from the British military establishment's refusal to adapt. Although luminaries like Capper saw the problem, the lesson had not been learned and the senior establishment dismissed the embryonic aviation development with some contempt.

Preston Watson's career in aviation and indeed, his life, took place within the period of these events and many of those mentioned (and many others have been omitted) impinged directly or indirectly upon his activities.

It is known, for instance that at some point he was in correspondence with the Wright brothers. Watson's own brother mentions the fact. Sir Garnet Wilson, Lord Provost of Dundee from 1940–46, was instrumental in negotiations to bring the National Cash Register Company's factory to Dundee. On subsequent visits to NCR headquarters in Dayton, Ohio, he was accommodated at the Dayton Guest House. This was situated at 7, Hawthorn Street, previously the home of the Wright family, where Orville's study is preserved. Sir Garnet reported [18] that Mr Allyn of NCR had heard Orville speaking of Preston Watson. The nature or date of the contact is not known.

Watson is likely to have known of Lillienthal's work and it is difficult to imagine he had not heard of Percy Pilcher. There is no mention of any mutual knowledge of aviation activity between Watson and Dunne although they were working for a time simultaneously in the same county – some thirty-five miles apart; nor was there any mention of knowledge of, or contact with, the Barnwell Brothers working a similar distance away, about the time that plane No2 was flying.

It is also likely that he had heard of Cody. Almost everyone had, in one respect or another, whether because of his showmanship, his competitiveness (he organised a race with a champion racing cyclist by riding a horse) [19] or his unusual exploits (he attempted to cross the English Channel from France in a small open boat towed by a kite of his own construction) [20]. It was more likely that Watson knew of him through stage appearances in his own Wild West Show. This was widely known and visited Dundee as Buffalo Bill's Wild West in Scotland on 19 and 20 August 1904 [21].

The era of air rallies, initiated by the French, came somewhat later. Among the best-known British ones were in 1910 at Lanark (in which Cody made an appearance) and Blackpool. There is no indication one way or the other as to whether Watson attended either of these as a spectator.

There is nothing to suggest that Watson had contact with Bleriot, but the engine of his second plane was essentially the same as that used by Bleriot in his celebrated Channel crossing in 1909. Watson is said to have had contact with Alberto Santos-Dumont from whom he allegedly purchased an engine somewhere between 1903 and 1906 (see Chapter IV).

Of course, one must not forget how far science had proceeded at the turn of the 19/20th century. The science of Fluid Mechanics was at a fairly advanced stage if one remembers that Bernoulli (1700–1782), mathematician and scientist, had developed his equation for calculating the relationship of fluids – which of course includes the air we breathe – between pressure, velocity and temperature. This is the basis of why an aeroplane flies. The camber of an aeroplane wing increases the distance that the air has to flow over the wing relative to the distance under the wing, thereby increasing the air's relative velocity. This reduces air pressure above the wing, producing a lifting force; causing the wing to rise and the aeroplane to fly. This is what Bernoulli's equation demonstrates. Later Venturi (1746–1822) developed it, and this same principle is used in the Venturi Tube for measuring the speed of aircraft.

All this basic scientific knowledge would be known to Watson and presumably other aviators. Watson had taken his knowledge of physics as far as he was able at school, but he needed to know more to design his aircraft, and this was presumably why he had further tuition under Professor Kuenan at University in Dundee.

The Wright brothers based their results on extensive wind tunnel research and empirical figures obtained from previous aviators, some of which they later abandoned. History does not reveal the approach that Watson took in devising his designs and the basis used for his calculations. He seemed to progress more rapidly from the point where he was experimenting with dead gulls, to the point where he was field-testing.

How much he was in touch with others in the field is also obscure, but he is very likely to have known the work of some of the earlier pioneers mentioned and have access to the proceedings of the Aero Club de France through the French published journals.

CHAPTER III

Preston Watson – The Man

Preston Albert Watson was born on 17 October 1880 at 24, Strawberry Bank in the West end of Dundee.

His grandfather, James Watson, son of James Watson was born in 1806. His second wife was Margaret Roger Preston, who was born in 1813. The couple lived at 17, Perth Road, Dundee (a site now occupied by the Duncan of Jordanstone College of Art). They had four children born between 1846 and 1852. Their third born, Thomas, was born in 1850. James, his father, died in 1861.

Thomas worked as an agent for a bakery supply firm in Bell's Court, Dundee. When the firm was threatened by financial difficulties, Thomas and his younger colleague, Joseph Philip bought out the owner and in 1873 they formed the company of Watson & Philip. They concentrated on high quality supplies to bakeries and the company thrived [22] until it was bought over by a larger concern in 1998. On 11 September 1878, he married Jane Yeaman and the family lived at 24, Strawberry Bank, Dundee. The Yeaman Family owned an engineering firm in the city and Jane's father, James Yeaman, was a prominent citizen, serving as a baillie on the City Council from 1861–64; and as Lord Provost from 12 July 1869 (when his predecessor William Hay resigned before the end of his term) until December 1871.

Thomas and Jane Watson had three children – James Yeaman Watson (25 July 1879); Preston Albert Watson (17 October 1880) and Euphemia Yeaman Watson (14 May 1887). At the time Euphemia was born, the family seems to have left Strawberry Bank and moved to 40, Magdalen Yard Road.

Preston was educated at Dundee High School and it emerges from accounts at the time and from photographs that he was a well-built, good-looking and popular boy who excelled in sports, particularly athletics and rugby. By the time he was eighteen, the family had moved to Balgowan on the Perth Road in Dundee. Preston was registered in the Newport (Fife) Rugby Club for the season 1898–99. He is described at that time as a 'fit young man of approximately six feet and twelve stone'.

John Bell Milne was a friend and fellow athlete and the two trained for, and competed in national and international events from 1898. Watson was the Scottish Long Jump and Hurdles champion in 1900; Milne was High Jump Champion for several years, starting in 1897.

Preston was a playing member of the Panmure Rugby Club when they were Scottish Champions and he played for Scotland against the Springboks. In 1905, he represented Scotland against Ireland in Gymnastics.

His further education is not clear but there is no record of him studying for a University degree or graduating from any institution of higher education.

However, Milne, in a letter to Watson's brother James written some fifty years later, recounts a meeting with Preston. The letter itself is undated but the rest of the correspondence would place it about 1953. He does not directly state the date of his meeting with Watson, but it could be inferred that it was as early as 1903 but probably not later than 1905/6. Milne stated in the letter that he was a student at University College, Dundee (UCD) in 1903. In the conversation it was revealed that Preston was also studying at UCD – his subject was physics, and his tutor Professor Keunan. The records [23] show that Watson enrolled as a student at UCD for the session 1900–01, thus introducing some doubt as to Milne's dating of events. The entry is un-numbered and does not specify the subjects studied. It seems likely from the record (or absence thereof) that he had left by the next session. It could be that he had arranged tuition on a more or less private basis from, or to join the lectures of the Professor solely for the purpose of furthering his aeronautical aspirations.

His brother James Y. states that Preston was clever at school and when he finished his education at Dundee high School, he "tried to get more information by attending classes at University College, Dundee (again setting Milne's dating somewhat earlier). There he studied higher physics..." J.Y. describes him making calculations about wind pressures and recounts a discussion where he was calculating wind drag between two mail trains passing at speed. He tackled the problem algebraically at a level well beyond J.Y.'s comprehension.

Professor Johannes Petrus Keunan was born in 1866 at Leiden in the Netherlands and attended University there, where his father was a prominent Professor of theology. He was appointed Professor of Physics at University College, Dundee (which was at that time a college of St Andrews University) from 1895–1907. His most significant work was on gaseous mixtures. During his tenure of office, Keunan had close connections with the Technical Institute of Dundee (colloquially known as Dundee Tech., and now the University of Abertay), acting as the UCD representative on the Institute's Board of studies for seven consecutive years between 1900 and 1907. There is no record of Watson having attended the Technical College.

Preston Watson's interest in heavier-than-air flight developed in his adolescent years and reports reflecting his thinking at that time depend mainly on his brother's accounts, recorded later. Like many others before him with similar interests, he watched the flight of birds. He was convinced that manned flight was a possibility and spent long spells in contemplation of the flight patterns of the seagulls which abound along the shores of the Tay estuary. The way in which these birds wheeled and turned using a dipping action of the wing was a significant observation on his part and one which was later incorporated into his designs. These deliberations would often divert his attention during group activities such as outdoors sports training and evoked much teasing from a sceptical peer group. He often sat on the seawall along the esplanade watching the seagulls, or launched models from the railway bridge at Ninewells. He even shot some seagulls to examine the wing structure and tail configurations. To some of them he attached lead weighting to study the effects of balance and trim.

It is not clear exactly when he decided to construct a manned machine and it seems likely that his interest simply grew steadily over his early adult years. James Allen in an excellent article in *The Scots Magazine* [24] describes how he found inscriptions on pews in Kinclaven Church. These were scratched or carved 'graffiti' presumably carried out in spells of boredom and are dated 1899 and 1900. One consists of the inscription "P.W. AVIATOR". Allen observes that pew 17 has Watson family inscriptions and pew 21 has Philip family initials. One of these reads B. M. Philip i.e. Beatrice Philip, daughter of Joseph.

The Philip family were domiciled in Perthshire and the Watson family in Dundee but the two patriarchs were not only business partners but good friends also and it is likely that the joint visits to Kinclaven church were at holiday times. It seems the Watsons were members of St Peters Church in Perth Road, Dundee.

In any event, Beatrice Philip and Preston Watson were married there on 19 December 1906, thus cementing further the bonds between the two families. At this time Preston was twenty-six and Beatrice twenty-four years of age. It is not clear exactly when Preston was taken into partnership in the Company, but it seems probable that it had happened a year or two before he married Beatrice. He was no doubt afforded a comfortable lifestyle thereby.

The young family lived at 'The Retreat' in the Western part of the Perth Road, Dundee. In due course the couple had two sons, Ronald and Harold.

Ronald served as a naval officer in the Second World War and lost his life in the Mediterranean in 1941. Harold graduated in medicine from St Andrews in 1938, obtained his Fellowship of the Royal College of Surgeons of Edinburgh in 1947 and worked as a surgeon in the Colonial Hospital in Trinidad, and as a naval surgeon. Harold's son also graduated in medicine.

Everything points to well integrated families on both sides. Certainly Thomas Watson supported his son in his experiments by providing inter alia financial help. But through family and business contacts, work space and probably engineering skills and resources were made available to him e.g. through the firm of Yeaman and Baggesen (Preston's mother was Euphemia Yeaman).

John Milne claims to have seen the first plane under construction on the premises of Preston's Yeaman cousin at Carolina Port. The date on which he saw it is not stated and neither is it clear that it was on or immediately after, the meeting referred to earlier. Thus it could be 1903 or even earlier, but not later than 1905/6. Milne wrote that it was about 1906 that Watson asked him to travel to France to bring back a French engine and fixed the date by remembering it as being before his examinations. (vide infra). It seems clear that for the whole of his adult life he invested a great deal of time and effort in the development of these machines. The premises for the building work were in another part of town and the trials took place at Errol and Forgandenny which would involve more travelling time then than it would now. He did own a motor car around 1903. This was a Deitrich which was chain driven and when the test flights took place he and J. Y. travelled out, usually on Sunday mornings.

How Beatrice, or indeed the other members of the family, viewed this does not clearly emerge but J.Y. wrote in 1955 that, as his wife, she had made Watson promise not to fly the second machine. (Either this meant not to 'test-fly' it, or Watson made no such undertaking, or he reneged on his promise for it seems certain that he did fly it); and later there are hints here and there that his father came round to thinking that he should reduce or give up flying activities, despite the fact that he gave substantial funding support to each machine.

During his early aircraft building, he enlisted the aid of a number of helpers. These seem to have been interested people – farm and estate workers enthusiastic youngsters, employees of the family firms, friends and members of his own family – to help with such tasks as maintaining and repairing and handling and stowing of the machine on the ground. The farmers on whose land he was operating seemed not simply to tolerate him but to actively support him with labour, storage space, ponies, and the like. Their families would bring cups of tea when work was going on and the general picture was of a young man who was very personable, engaged in an activity that excited a lot of general interest. One of the helpers, John Logie, thirteen years old at the time, helped with the second aircraft. In a letter to J.Y. he later described Preston thus: "He was a wonderful and daring man" "I think he was a very cool and courageous man. He was very determined, and never lost his temper no matter what happened".

When war broke out in 1914, Preston Watson volunteered for active service. It appears that like almost all young men of the time he had strong patriotic sentiments and was willing to enter the conflict. (His brother-in-law William Philip was a conscientious objector, but no doubt a strong patriot too). But more than this, Watson obviously felt that his experience of building flying machines of his own meant that he had something specific to offer which was of great potential value to his country.

He had held a commission as a volunteer in the Royal Forfarshire Yeomanry and his expectations on enlistment were that he would be granted a commission in the newly formed Royal Flying Corps, which was not yet two years old at the time. Viewed from his own perspective at the time (and from that of the reader more than a century later) this would seem reasonable. It proved to be less easy.

In pursuit of these aims he travelled to London armed with an introduction to a Major Houston, no doubt from a contact in Dundee or the Forfarshire Yeomanry. From Houston he was referred to a Major Merrinden. Watson's account of the interview with Merrinden in the following letter is illuminating. The fact that Watson had not been to a public school seemed to be regarded by Merrinden as a major impediment. Ignoring Watson's rare and valuable flying experience, he pronounced him too old to be a pilot at thirty-four years of age (the actual maximum age at that time was forty years) and suggested he might try for a commission at Farnborough.

> Dearest,
>
> I got your sweet letter this morning. Never mind Bea, someday I'll come back and it will be fine.
>
> I don't know whether to bother you with all the interviews I have but the result is that they consider me too old to be a pilot, or at least they say so. The sort of influence I have is not close enough. For instance I heard Major Houston phoning to Major Merrinden and Major H. just said "This man Watson is recommended to me by a friend of mine. I don't know anything about him but you might see him." I believe I would be better without that kind of thing. And Merrinden's questions to me: Have you been to a public school simply gives one the pip.

It is not nice being refused a job. There is one card I may play yet but I don't want to if possible.

It is this:

Ashby Mitchell introduced me to two men, one was Underhill a solicitor who knew Major Houston. The other is a Mr Jouques. Jouques is such a big pot that he says he can get me a commission by raising his little finger figuratively. But he makes conditions. He wants me to take a commission at the Aircraft Factory or in the Aircraft Inspection department and this for a special reason. I am going to see Jouques tonight because his proposal comes more into line with what Major Merrinden said to me on Saturday. Major M. said I recommend you to go to Farnboro and apply to Col. Fulton for a commission in some sort of connection with the experimental work at the Aircraft Factory.

I have this morning just got a letter from Major Merrinden saying my age is over the regulation age for a pilot.

Well this man Jouques has been building Aeroplanes in Russia and was introduced to the War Office by the Russian Government. His wife is a daughter of the Lord Mayor of London, not the present lord mayor of course. She is also a niece of Lord Kitchener's.

I was at Jouques house the other night and met there the Director of Aircraft who occupies the same position in aeroplanes as Churchill does with navy matters so you see this is high society. By the way the blaze of diamonds was simply immense. I must say I thought it rather vulgar. Their house is in Cavendish Square and is most magnificent.

Well, to proceed with the tale!

Jouques is building six aeroplanes for the war office to be ready in 8 weeks under penalty of £5,000 which he has paid down

On Saturday he accepted an order for another fifty aeroplanes to be ready in six months under penalty of £10,000 which he paying (sic) down today. He has bought a new factory.

Jouques has in his employment a Mr Posener who was Chief of the Aircraft Inspection Dept., that is chief next to the commissioned officers. Posener was in fact the man who did the inspection work. Mr Jouques offered Posener £15 a week to become his employee and help him build the aeroplanes and Posener accepted and is now in Jouques' employment.

Of course Posener is a great help to Jouques because he knows all the work and in fact if he passes it, then the Government just passes it too. So Mr Jouques did a pretty cunning thing in employing Posener. Besides Posener is a pretty good man as he was appointed to Aircraft Inspection Dept by examination, he being first out of 300 in the theory and practice of aeroplanes.

Well I showed Jouques the description of my machine, Mitchell had mentioned it to him, and immediately Jouques is on it like knife and called in Posener who went off and spent seven hours over it. He evidently reported something good to Jouques for this is what took place.

Jouques said "Look here you must demonstrate this to the War Office" I said "No thanks". He said "If you do I will get you a commission" I said what exactly did you mean. He said "You will pay the expense of this demonstration and if successful I will pay you £25 royalty on each machine that I make and £40 per machine of this kind built by anyone else".

I went home and wrote him as follows.

Mr Jouques

Dear Sir

I thank you for your kind offer which I regret I cannot accept.

Will you please ask Mr Posener to return now the photographs, etc.

As you will probably be much engaged with your other guests on Saturday I ask you to excuse me on that evening.

Yours truly

P.A.W.

Next day my landlady told me he telephoned three times asking for me and wanting me to telephone him. I did so and he said Mr Watson do come on Saturday night, I have some things to speak to you about. Well I went and he hadn't any other things to speak about but he had spoken to Posener again and also to the Government Commission and here was his new proposal.

I shall send my wife to Lord Kitchener who will get you a commission in the Royal Flying Corps at first in the capacity of Aircraft Inspection Officer, your duty being the supervision of the demonstration of your machine. I shall send a man to Dundee to pack it up and bring it to Farnboro at my expense. I shall at my expense put new skids and springs and brakes on it and employ a pilot to fly it. It will cost you nothing. I shall insure it against accident and if it is not successful I shall put it back in Dundee in the same condition as it is now. If it is successful I shall pay you the royalties of £25 and £40 or £1,000 plump down.

You will get your desire to have a commission. I get what I want; the Government will have your services in the position they recommend you to occupy and will have the chance of benefiting by your invention.

He messed around and gave me coffee and Russian cigarettes and drinks as if I had been the Prince of Wales which I must say was not disagreeable after what I have had at the War Office and Admiralty for the last two weeks. Do you know I believe I have had about 20 interviews in that time?.

I am now swithering. Mr Jouques is away to see Kitchener at 11am today. I am pleased about one thing, the despised invention is not despised here. The reason I have written this is to show that. But I know the pater doesn't want me to have anything more to do with it, and which by agreeing to Jouques conditions I will get a commission in the RFC and won't come back to Dundee having failed, still it's not altogether serving my country because I stand to make money out of it too. Perhaps you will manage to go along to Balgowan and see what the pater thinks?

One good thing would be that I would get home for a few days, because while Jouques sends men to Dundee to pack up the machine, it would be more reasonable for me to do it and he pay me instead of the men.

In any case I'm not worrying myself in the least because in time of war, one expects all sorts of extraordinary things. So don't worry either.

Love from

Your Pres

There's no doubt about Jouques truthfulness, I have seen his aeroplane factory and he showed me a duplicate of a cheque for £4,050 which he paid for 50 tons of special aluminium because he wanted a cwt and the man wouldn't sell less than 50 tons! Of course he sold the rest at a profit afterwards!

I enclose a charity concert programme with Jouques name on it showing who he is mixed up with.

He is quite direct about being offended by the lack of courtesy and the obstructive attitudes of the authorities. He was an established business man by this stage in his career and it is clear that he was reserved, indeed cautious, about the proposals being put to him, but the letter gives the impression that he was judging the London scene by the standards of loyalty, transparency and integrity that he enjoyed in the 'smaller pond' of Dundee commercial circles.

The letter also illustrates how keen and determined he was in his pursuit of aviation.

He is almost justifying this to his wife and attempting to enlist her aid in persuading his father to approve. But it suggests that he felt that she was unhappy or unwilling with her support in the venture and that Watson was attempting to assuage her misgivings. It is not entirely clear how much guilt he felt. The situation must have been difficult for him and it is difficult now to imagine the attitudes and mores of the early months of the First World War.

The letter produces a fascinating cameo of some of the practices of procurement of war ordinance in a military establishment undergoing a transition from patronage, ceremony and pomp to an efficient and fit-for-purpose army; and a transition from massed foot and cavalry assaults to a war fought with mechanized land forces, submarines and aeroplanes.

The route taken by Watson to his commission in the armed forces is not clear. From his letter home dated 1 February 1915, he was obviously hoping for a commission in the Royal Flying Corps (RFC), established in 1913. Merrinden had stated (incorrectly) that he was too old to be a pilot and suggested that he might try Farnborough for, presumably, a non-flying appointment relating to aircraft. Jouques promised to secure him a commission via Lord Kitchener, the Minister of War at the time, and that would be arranged through Colonel Fulton at the Aircraft Factory. This would be in the capacity of an aircraft inspector, the implication being that he would be inspector of his own machine. The proviso was that Watson would co-operate in a demonstration of the rocking wing (see Chapter IV) to the War Department.

There is no record of the proposed demonstration of the rocking wing to the War Department ever taking place.

Having been told he was too old to be a pilot, he undertook flying training at his own expense at Hendon, obtaining his Royal Aero Club licence (No 1117) on 16 March 1915, training on a Beatty-Wright trainer [25, 26]. The RNAS seems to have been a much less formal organization than the army and had to compete with the RFC for pilots (and presumably planes).

Whatever route it was, events must have moved very rapidly after his meeting with Jouques in February, because he had completed his training as a pilot, obtained his licence, visited his home in Scotland, and been commissioned in the Royal Naval Air Service by 29 April.

Jouques seems to have been quite influential with many accessible contacts in high places. He had told Watson that his wife was Lord Kitchener's niece and was the daughter of an ex-Lord Mayor of London. Both of these statements seem to be less than accurate, but he did seem to have ready access to Lord Kitchener if he was able to go off to discuss with him the morning after meeting Watson. It could have been that the account of these particular connections related to Watson were either misheard or misinterpreted by him when he was a guest among a number of others at Jouques' home.

Leo Anatole Jouques was born a Russian citizen in the city of Vilna in Lithuania in 1864. The family was Jewish. His parents, David and Rebecca Jouques both died in Vilna. Leo graduated with honours in law from the University of Warsaw and continued his studies at l'Ecole des Hautes Etudes in Paris. He practiced as a barrister in Vilna until 1889 after which he moved back to Paris until 1892. After a year in Chicago, he again returned to Paris and moved to England in 1909.

There he married Violet, daughter of Viscount Walter Radcliffe Horncastle. Walter Horncastle had been made a viscount by King Don Carlos I of Portugal in 1895. He was a member of the Corporation of the City of London, becoming Chief Commoner in 1903, and was the first Mayor of Hackney [27]. (It was probably this piece of information that had been misinterpreted by Watson as meaning Lord Mayor of London; but by an odd co-incidence, Jouques at the time was living in rented property at 6, New Cavendish Street. His landlord was Sir William Henry Dunn who was to become Lord Mayor of London the year after Watson's visit!). There does not seem to be any family connection to Lord Kitchener. Violet Horncastle was married to Leo Jouques on 31 March 1909 in the West London Synagogue, suggesting that the Horncastle family were also Jewish.

Jouques worked to great effect on behalf of British industry, brokering sales of British made munitions to Russia in competition with the Krupps industries in Germany. He also travelled in Russia and Eastern Europe for the Marconi Company and succeeded in persuading them to set up Marconi radio stations, again in competition with German industry.

At the outbreak of war he set up a business to manufacture aircraft on government contracts. Jouques Aviation Works, Willesden [28] later joined the Society of British Aircraft Constructors (SBAC). This was a large consortium set up in March 1916 to include a variety of companies of different sizes and activites, some subsequently very well-known, to deal collectively with the Government. The initial premises of Jouques Aviation were in Willesden, but after a short time they moved to larger premises in King's Road, St. Pancras.

Watson, in his letter home quotes Jouques as mentioning two contracts – one for six aircraft and another for fifty. Subsequent records reveal that this was exactly what was delivered – a total of fifty-six planes – as well as parts for some 150 more. The planes produced were predominantly BE2c s but some other models – BE2b s and BE2e s were also manufactured. These were 2-seater biplanes designed by Geoffrey de Havilland in the

Royal Aircraft Factory in 1912 and put out to a number of manufacturers for wartime production. They were very stable which made them suitable for bombing, night fighting and reconnaissance, but very vulnerable to the faster and more nimble German fighters. The original version was the BE2 and production continued, with a number of modifications (2a, b, c, d, e) throughout the war. Full details are recorded by Munson [29], including illustration of a Jouques-built BE2b.

In 1919 Jouques, who had donated a considerable sum to the war effort, applied for, and was granted, British Nationality. In the same year he wrote to Major Lionel de Rothschild offering a donation of $5,000 for the foundation of a Jewish university in Haifa [30].

Watson's plane No3 went to Buc, near Paris, in March 1914 and was there for four months. There is no account of its return, but it was presumably returned to Dundee in August of that year, shortly before the declaration of war. This would be the machine proposed for the demon-stration to the War Department, although it seems unlikely that this ever took place.

In that context, it is interesting that part of Jouques' offer included renewing the skids on the machine (Plane No3 was originally fitted with wheels) but perhaps Jouques and Posener, his technical assessor, had only been shown plans of plane No1 as they appeared on the patent documents. There is a photograph of plane No3 in flight on which can just be distinguished a set of skids in addition to the wheels. J.Y.'s note indicates that this was at Buc.

The Royal Naval Air Service (RNAS) was created by the Admiralty from what had previously been the Naval Wing of the Royal Flying Corps (RFC) possibly without their full agreement. The RFC resulted from a committee recommendation to form a flying corps with a military and a naval wing. The latter was for scouting and coastal reconnaissance but after formation of the RNAS in July 1914, it was given the task of defence of the Country while the army was engaged off British soil. The fear was of attack by submarines and zeppelins. The main base for the RNAS was at Eastchurch and it was there that Watson underwent his training for his commission.

In July 1914 they had 130 officers, thirty-nine land planes, fifty-two seaplanes and seven airships for the task. War was threatening and presumably there was haste to increase strength so that once Watson's availability was realised in the appropriate quarter, it is likely they would welcome him for flying training and a commission.

Preston Albert Watson was commissioned Flight Sub Lieutenant (temp) on 29 April 1915.

CHAPTER IV

The Preston Watson Planes

Preston Watson designed and built three aircraft.

PLANE No 1

The bulk of evidence from those involved in its development and eye-witnesses indicates that Plane No1 was under construction in 1903. In this chapter it was thought that it might be helpful to set out the substance of these accounts as collected by J.Y. and to go into some detail about these sources in an attempt to bring into some degree of focus the individuals concerned. This allows a 'story' to be constructed for the building of plane No1, in much the same way as J.Y. must have done, along with his own recollections while preparing his initial claim.

This claim was rebuffed (see Chapter VI) and finally J.Y. presented a dossier of his findings to the McManus Gallery in Dundee. At the start of this he has composed a summing up in which he writes "I do believe from the evidence contained in this album that 1903 was the year that both of us, after many attempts, were airborne for quite a number of times for a few minutes".

John Bell Milne was a student at University College, Dundee in 1903–1904. He was described as studying dentistry, but Dundee had no dental school at that time. It may be that he was studying basic medical sciences. He continued his studies in Edinburgh and went on to have a distinguished career in dentistry, with a position at St George's Hospital in London, from which he retired in 1947.

In an undated letter from a Dundee address (probably written in the early fifties) he mentions that he was a student at UCD in 1903 when he learned that Watson was having tuition from Prof. Keunan (that tuition was 1900–1901).

He saw the plane (the implication is that he saw it in 1903, but that is not stated unambiguously) in the workshop at Carolina Port. This was part of Dundee's dockland and presumably on the premises of Yeaman and Baggesen, Watson's grandfather's engineering firm, managed by his cousin. Milne describes the plane thus: "The framework was that of a glider. There were no wheels and no engine" and he goes on to state "Unfortunately I was not at Errol thus I cannot say first hand what happened there".

David Crighton worked at Durkie's Stobswell Bakery and is quoted by his daughter as recounting how Watson used to bring lengths of bamboo into the bakery to be placed in a 'cold' oven to warm them for bending and shaping. When this occurred was not reported, but he remembered Watson coming in at a later date, jubilant because he had been airborne. David Crighton dated that event at 1904–1905.

In the first week of September 1902, Henry Band returned home to Errol from South Africa, where he had served in the Boer War. He remained in the district, unemployed, until August 1903 when he left to take up work elsewhere. During this time he used to watch the plane flying; and indicated that he would be willing to swear on oath to that effect. He was absolutely sure of these dates as he married shortly before he left. The marriage certificate is dated 5 June 1903.

He compared the sound of the engine to the sound of machine-gun fire and is not likely to have confused this with the sound of the launch weights falling.

Alec Robertson gave J.Y. an account of his involvement which is recorded in the letter J.Y. sent to Lord Brabazon (see Chapter VI), in which he says, "sixty-three year old Alec Robertson 49, Castle Street, Broughty Ferry who, when he was a schoolboy of thirteen years of age, towed Preston's first machine by a pony to a cart shed at Leys farm, Errol, a distance of about a mile. He remembers assisting Preston to dismantle part of the plane to make it more suitable for towing, and so that it could be housed in the cart shed. He handled the pony that was borrowed from Mr Melville, the farmer at Leys farm. All this took place before he left school. He left at thirteen and went to work on the farm where his father was grieve". This would set the date at 1903/4.

David Urquhart, who worked in the motor trade for the Thomas Shaw company, was the founder-member and secretary of the newly formed Dundee Aero Club in 1908 "by which time plane No2 was flying."

He remembers that the club was given permission to reclaim the by-then derelict Plane No1. He wrote to J.Y. to say "At this time (1908) the first plane was lying at some farm all dismantled and in pieces and Preston delivered it to us by truck in that condition to be assembled and used as a glider…" "as a glider without the engine it performed wonderfully". He states that he had a recollection"that would definitely establish the exact date of the 1903 aeroplane. I had this in my possession in the form of the plans which were dated and I am confident that is why we came to always refer to it as the 1903 plane". He noted that at one time the club also had the engine, which they believed to be an ex-balloon engine and Urquhart states that it was a Dutheil-Chalmers.

It was subsequently lost and never again came to light. This is – at least in part – corroborated by W.R. Gibbs [31] who described in a lecture a 10hp Dutheil-Chalmers 2cylinder engine which was light enough for him to carry about during the restoration of the plane in 1908. However, photographs show that there were eventually two such units on plane No1.

Urquhart married and left Dundee to work for Rossleigh in Edinburgh in 1910. He was described by J.Y. as being very knowledgeable on the aerodynamics of the time and worked closely with Watson on the second plane. He returned to Dundee in 1912 to deliver a lecture on the subject to the Literary Society of Broughty Ferry.

John Gourlay, whose father was head groom at Rossie, Forgandenny and who helped with the launching mechanism during tests at the Haughs, near Forgandenny station, wrote to J.Y. to say that his mother used to take his sister (born 1899) to watch the plane flying. This was before she was of school age, placing the year as 1904–5.

One July morning in 1903 two lads, twelve-year-old Alec Patterson and his friend Earnest Bruce saw a strange contraption being drawn by a pony along the roads leading to Leys Farm (near Errol, Perthshire), where it was put in one of the sheds.

The next day the boys were there again to see it re-emerging as they had been told that an attempt was to be made to make it fly. The machine was maroon with the addition of some overnight decoration with hen droppings.

The machine was taken out and tethered by a rope from its tail to a strainer post. The boys were asked if they would release the tether by untying it or cutting it when the signal was given. This was given when one of the two men was seated on the machine and the engine was running at full revs.

When the tether was released – by cutting, as it turned out – Patterson saw the plane leap forward but it did not leave the ground. The two men were, of course, the Watson Brothers who persisted with some further attempts, the best of which resulted in what could best be described as exaggerated bounces. After this, they changed places and with J.Y. as pilot the plane left the ground for 100–200 yard hops.

On one of these attempts, Alex Patterson observed J.Y. rising to first storey height and then the machine nosed down, hit the ground and then somersaulted forward, throwing J.Y. clear but breaking the propellor.

"Preston spun the propellor and we cut the tethering rope. Away she bounded with us running barefoot in its wake. After a hundred yards or so the machine suddenly lifted, and rose to about the height of the farm buildings".

Patterson also mentioned that it was a front engine and, as boys, they were forbidden to stand in front of the machine. He was quite sure that the month was July because the two boys saw Melville, the farmer, preparing a mare for the Errol Games the following Wednesday and the games were always in July. He remembered, too, the brand of lemonade that the Watson's gave them at lunchtimes (Ballingall's) [32].

The amount of detail given in this article is remarkable for recollections sixty years later; but he was a detective by training and had more than likely rehearsed the story on a number of occasions. Whether he also revised it has to remain a matter of opinion. The account of the crash does sound very similar to what J.Y. describes as happening to him in plane No2 (although on that occasion J.Y. describes climbing to about 200 feet). On the other hand, many of the less successful hops of those days would be likely to end up the same way!

Kerr B. Sturrock ran the engineering firm of Sturrock Patent Bridge and Engineering Company in Dundee. It was he who, in person, designed and made the propellers for the plane. He remembers that Watson was working on his first machine "long before" he (KBS) was married on 15 September 1905. He stated that at that stage the first machine "was off the ground but would not stay".

Sturrock's testimony is also interesting in relation to the early engines as will be seen. But Kerr B. Sturrock deserves special mention – indeed he deserves a special place in the history of aviation.

He died in 1957 in Dundee and his obituary [33] reads as follows:

One of Dundee's leading businessmen Mr Kerr B. Sturrock, seventy-four-year old principal of the joiner's and shop-fitter's firm that bears his name, died at the Royal Infirmary after a two month illness.

He pioneered in aircraft propellor construction, and a former engineering product of his firm was adopted by steamships all over the world.

Mr Sturrock, who celebrated his golden wedding two years ago lived at 17, Dalgleish Road.

Born at Kerrystonebank, near Murroes (seven miles North of Dundee), he served his apprenticeship as a joiner and patternmaker in Dundee, and in 1903 became associated with a brother, Robert, in the Sturrock Patent Bridge and Engineering Co. in Peter Street.

The firm's "bridges" replaced the brick backs in steamship's coal furnaces and ignited the smokey waste gases. This saved fuel and cut out the clouds of black smoke that belched from most steamers' funnels. Shipping companies took up the idea and agents were appointed in many countries.

The patent rights were sold to a Greenock company in 1936. Shop-fitting became Mr Sturrock's main business and the firm's activities – as Kerr B. Sturrock & Sons, 25 Baltic Street – now extended far beyond Dundee.

AIR PROPELLERS

Early in his career Mr Sturrock succeeded after many attempts in producing an aircraft propellor of wood for Dundee air pioneer Preston Watson.

He had no previous examples to guide him. But the method was so successful that it was generally adopted, and was only dropped during the last war with the introduction of light alloy propellers.

Mr Sturrock was a long service elder of Wallacetown Church. He was a member of Masonic Lodge F & K.

He is survived by his wife and two sons. The business is carried on by Richard. Gordon is a teacher at the Trades College.

(Kerr B. Sturrock died 14 October 1957, three days before James Y. Watson).

Sturrock started by constructing the propellers of yellow pine which "proved useless". Attempts with other woods proved to be no better. About his twelfth attempt he writes,

> In this case I jointed Australian walnut 1/8th" thick and laid them out of line so that there was a certain amount of cross grain. This proved the best construction and was adopted by all other makers and was in use until well on in the Second World War and was then displaced by a strong and very light alloy.

These laminated propellers were used on the famous Mosquito twin-engined pathfinder/fighter bomber in World War II and can still be found in use on some light aircraft in the 21st century.

Schulman [34] attributes the invention of the laminated propeller to the Aerial Experiment Association led by Alexander Graham Bell – and in particular to Glen Curtiss circa 1908. Although Kerr Sturrock had devised them prior to 1905 it is most unlikely that either knew of the other's work. In an era of intense, and often competitive, interest in becoming airborne, the slower communications and the separation between Europe and America, it is reasonable to assume that both men arrived at the same conclusion independently. Sturrock was first but there was little other local publicity, whereas Curtiss was much in the public eye in America and went on to found an aircraft construction company of his own.

Sturrock describes how a second engine was coupled to the first and states that a lot of alteration to the plane and propellor was required, although he does not go into any detail.

THE PLANE'S FRAMEWORK

The plane's framework was constructed from bamboo with four long bamboos running the length of the plane and attaching a tailplane of box-kite construction which pivoted around a transverse axis. The central construction was in the form of a tapering tower with four lengths of bamboo, two forward and two aft of the central pilot seat on the port side; and the same on the starboard side. Each port side length met its corresponding starboard length at the top and was attached at the foot to a length of wood which formed the basis of the skid. The skid was soled with metal and this sole was extended forwards and curved upwards (like a sledge runner) and was joined to its mate at the front by a transverse wooden bar, and linked by a single central spar to a second transverse bar fixed across the forward pair of bamboos of the tower at about the level of the pilot's seat.

The main wings were positioned as for a low-wing monoplane and were wire-braced to the top of the tower. This system was used in principle in some of the very early designs e.g. Henson's ariel steam carriage of 1842 which used wire bracing from a central post; Felix du Temple's innovative design of 1874 and Lillienthal's monoplane hang glider of 1894. (see labelled diagrams)

However, Watson's design was simple but very astute. The central tower in Plane No1 was not simply a device to brace the wings. Being of triangular section (when viewed head-on), and by extending right down through to a broad base at ground level, it introduced intrinsic lateral stability on contact with the ground – very important for take-off and landing.

The triangular shape of the tower and the triangular shape of the fuselage at right angles to it would certainly limit a tendency to distort in a structure that was so lightly built. The strength at the top of the triangle provided a good pivoting point for the single top wing, which did not require bracing.

Control was effected by means of a 'stick'. This was attached to the underside of the top wing so that side-to-side to side movement of the stick tilted the top wing to give lateral control i.e. control in roll and correction of side-slip; and to allow steering. The stick was also connected to the tail planes so that fore-and-aft movement of the stick tilted the tailplanes to act as elevators. Thus the pilot had a single control for these functions which he could operate with one hand. There was none of the weight shifting, pelvic movements or shoulder harness required by glider pilots up to that time and there were no cumbersome mechanisms for wing-warping.

These features were definitely in the design of plane No1, as can be seen in photographs and patent diagrams. In all these respects Watson was well ahead of his time (as were Sturrock's laminated propellers). The tilting top wing was only bettered when the aileron was used by Curtiss on his June Bug plane of 1907 but even then the ailerons were controlled by a yoke fitted to the pilot's shoulders. The concept of the aileron had been put forward by Boulton in 1868 and was the subject of experiment on a glider by Robert Esnault-Pelterie in 1904, [35] but according to Schulman [36], it was Curtiss who first used them effectively and set them on course to be the standard method of lateral control in aeronautics, despite the claim of the Wright brothers to the contrary. Their method of wing-warping did persist for some years and was used by some aircraft constructors such as Bleriot. But even the Wright Brothers' own Company discarded the system in favour of the aileron in 1915. It is ironic that for years Orville Wright pursued Curtiss though the law courts for infringement of patent; and not without initial success.

The wings of plane No1 were cambered with an upper surface that described a good aerofoil shape, but as far as can be seen there was not a plane undersurface. The wing span was estimated by Smith [37] to be 7.6 metres (25ft).

Working from the patent illustrations and the models held in the McManus Gallery in Dundee, the total wing area for all three wings would work out at 15.5 sq metres and the area of the two horizontal surfaces (plane) of the box-kite tail would be 1.8 sq. metres. These are, of course, estimates. (for comparison the plane surfaces of a Cessna 152 amount to 43 sq metres. The little CA-25N "Gazelle" has a wing area of 11.6 sq metres; a Rotax engine producing (max) 80hp at 5,500 rpm; and has a wooden propellor).

Several authors have described the launching mechanism used by Watson with, in some cases (e.g. *The Meccano Magazine*, 1957 [38]) a diagram showing a cluster of agricultural weights and a blacksmith's anvil. These were suspended by a rope over a pulley attached to the stout limb of a tree. This rope was run through a pulley at the base of the tree and thence forward to a snatch-block and then run back to a cradle on which the plane rested .The diagrams are likely to be based on a drawing made by Urquhart.

James Elder, who lived in Errol at the time, was later able to take J.Y. back to the stump of the tree that had been used on Leys Farm. John Gourlay recognized, from a photograph published in the *Courier*, the area where the launches were made at Rossie and he and Smith [37] were able to identify the tree still standing in the 1990s. He also had a shed in the steading partially surviving one hundred years later.

There was a release mechanism apparently operated by the pilot to allow the weights to fall thus propelling the cradle and plane forwards. The technique was to open the engine up to full power and then release the weights. Some descriptions reveal that initially the mechanism was lubricated with lard (perhaps from the family business!). Gourlay describes how one day Watson turned up in his Deitrich motor car and the idea was put forward to use the graphite used on the car's drive chains instead. This proved to be much better.

Urquhart provided further detail in a letter to J.Y. in 1954. "The plane simply rested on the cradle which had a shoulder at the back so that the plane could not slip off. The only slip-hook used was the one the pilot released to enable the weight to drop. When the plane reached the end of the rail it simply shot off. Their engine was running full blast at that moment and that is how all the initial flights were made". The cradle slid along a rail some thirty yards long; but would only act as far as the weights could fall from tree to ground.

According to Urquhart the system was discontinued and not used for plane No2.

The various references made to the location of the trials are confusing. The four sites mentioned are:

a) Muirhouses and Leys farms at Errol
b) 'Dovecot Park' at Rossie near Forgandenny
c) 'The Haughs' at Forgandenny Station
d) Belmont, near Meigle outside Dundee

The last of these can reasonably be dismissed as a confusion in an article by J.D. Leslie [39], arising from the fact that the owner of Dovecot Park, J. Heriot Bell, had his primary residence in Perth Road, Dundee entitled 'Belmont'* and this name had been mentioned in a letter from J.B. Milne to J.Y.

There is little doubt that the earliest test site was Errol, in the Carse of Gowrie about six miles from Perth and fifteen from Dundee. The sites were on the farms of Leys and Muirhouses, situated respectively North-West and North-East of, and adjacent to, the area that was to become Errol Aerodrome. Over a century later, in the early 21st Century, this was still identifiable as an airfield although no longer used except as markers on the approach to Dundee Airport and boundary mark for the approach to Perth Aerodrome at Scone.

J.B. Milne wrote "Some difficulty arose and it was decided to remove the machine from Errol". It was Milne who brokered the arrangement with Heriot Bell to allow trials to continue at his country estate at Forgandenny. A thirty acre parkland (Dovecot Park) was put at Watson's disposal. J. Heriot Bell was John Bell Milne's uncle. Milne also mentioned "There was a good deal of secrecy about the building of the first aeroplane..."

---

* This property was very close to 17 Perth Road where Thomas Watson's parents had lived and his mother and brother were still living at the time. The name is perpetuated in the student residence, Belmont Hall, after the area was redeveloped.

In a later correspondence with Charles Gibbs-Smith (*vide infra*), J.Y. clarified both these points. There was a problem with the early trials being too much in the public eye. Estate and farm workers, schoolboys and the local inhabitants were keen observers. The reason for the secrecy was that up to that point no patent applications had been lodged. It is also possible that there was some realisation of military potential. John Harris noted that Watson was always on his own when he came to Forgandenny and presumably only had the help of trusted estate employees to move the planes.

There is no evidence whatsoever of mutual awareness between Watson and Dunne although they must have overlapped in time and were only working some forty miles apart. Capper was known to have an appreciation of the military potential and moved operations from Farnborough to Blair Atholl entirely because of the secrecy issue.

The date of removal from Errol is not clear although it is likely that it was around 1905. The catapult launch was used at Forgandenny and Plane No1 was photographed there in 1906. It is difficult to pin down when Watson was at each of the sites. The clues would indicate that he started in Errol about 1903, and then moved to Forgandenny around 1905, firstly at Dovecot Park and then 'the Haughs'. Plane No1 was off the ground at Forgandenny but required the catapult launch. Even then it was difficult, or impossible, to keep it airborne for anything more than short hops of a few seconds. It is possible that plane No2 was started on test flights in 1908 (*but see chapter VI*) and was evidently back in Errol and flying in a convincing and controlled manner, probably by 1909/10.

In 1908, plane No1 was donated to the newly formed Dundee Aero Club and the engine was removed. A pair of bamboos was removed from the forward component of the tower (this was confirmed by J.Y.) and rotten sections were restored with bamboo obtained from a firm in London, but the extra bracing that Watson had added as reinforced support for the engine remained. The restored glider had two wheels on a single metal axle as well as the skids, but Urquhart describes how this was rusty and had to be replaced, suggesting that it may have been used at Errol to facilitate transportation on the ground.

Members of the club wheeled the newly restored aircraft through the streets of Dundee to the Tay Ferries and took it across to St. Fort near Newport in Fife where it was successfully flown as a glider. Urquhart remembers pushing it downhill for its first take-off.

According to the *Blairgowrie Advertiser* [40] (Blairgowrie was J.Y.'s hometown) the plane was ultimately returned to Dundee and dismantled; but the date and circumstances are not recorded.

The chronology of the No1 machine would be greatly helped if the engines fitted to it were accurately known. Unfortunately, there is a discrepancy in the available information about them which is unlikely to be definitively resolved. The best that can be done is to review the available evidence – and its reliability; and to look at contingent circumstantial facts in order to arrive at a 'best fit' view.

One of the more helpful accounts is that given by Kerr Sturrock. He states that the first machine had a 10hp French engine which was later coupled to another to give greater power. He refers to the French engine as a de Dion and describes it smashing a fair few of his early attempts at making a propellor.

If this were the case, there would, at this time, be several candidates. A number of correspondents refer to the fact that an engine was purchased from Santos-Dumont, the well-known Brazilian airship constructor and aviator who worked in France. This probably originates from J.B. Milne's statement that Watson was travelling to Paris to collect it, and J.Y. refers to it as an 'old balloon engine'. The most likely candidate engines to fit these facts are either:

1 The de Dion engine used by Santos-Dumont in 1898

This had been removed from his motor tricycle and after some experimentation, he redesigned a "double de Dion" which was constructed under his direction by the motor engineer Albert Chapin in Paris to Santos-Dumont's specifications. This had a very unusual configuration in which part of a second engine was mounted above the first and is described by Wykeham [15] as having "the top piston linked to the lower by a rigid connecting rod, through two gas-tight joints This engine was evidently running in 1898 and he re-fitted it to his motor tricycle. It proved its worth by keeping up with the front runners in the Paris – Amsterdam race. (His motor tricycle was not eligible to enter officially and he did not complete the course, having satisfied himself on the performance of his engine).

This engine was fitted to airship No1 which made her maiden voyage in September 1898. As Santos-Dumont no longer had a functioning motor vehicle, he purchased a 6hp Panhard which he changed shortly thereafter for a battery driven electric brougham.

The airship engine was used in the ill-fated airship No2 (May 1899) and airship No3 which first flew in November 1899. Therefore, this engine could have been available for Watson to purchase from late 1900 on.

2 The engine first fitted to number 4

Airship number four first flew in August 1900. It was initially fitted with an engine of similar design to the previous 'Double De Dion', comprising two cylinders developing 7 hp, and the older version was then discarded.

Santos-Dumont's hope for this airship was to use it to win the prize of 100,000 francs put up by M. Deutsch de la Meurthe for a flight around the Eiffel tower. He found that the engine he had first fitted did not guarantee enough speed, so it was changed, after only a few weeks, for another engine of his own design, this time a four cylinder air-cooled engine which he had built by Buchet. This engine was heavy enough to require a significant enlargement of the gas envelope, lengthening the airship by 14 feet.

It seems quite possible that the original No4 engine would be available for purchase from late 1900 onwards.

Other possible sources of a De Dion engine are that Watson may simply have purchased one.

The Argyll Motor Company was using De Dion engines from 1900 until about 1904. Their premises were in Alexandria, near Glasgow and it is conceivable that Watson could have obtained one from them.

3 In a letter dated 26 July 1957, J.Y. wrote in reply to one of the letters from Charles Gibbs-Smith:

Enclosed are old photographs, one of the De Dion Car, the engine of which Preston may have used in his attempts, but not in 1903. I make no claims that the "machine" that Preston used in 1903 at Errol was a powered machine, but it was a "rocker", of that I am certain.

The sentence construction here does not adequately clarify whether the photograph was merely an example of the make of car, or whether it was a car that was available to Watson, the engine of which was removed for use in early attempts with plane No1, as Sturrock's testimony seems to suggest.

Lining up the dates is not easy. J.B. Milne, in his letter to J.Y., mentioned that he met Watson in 1903. He offered to negotiate with the owner of land at Forgandenny (Mr J. Heriot Bell of Belmont, J.B. Milne's uncle) for trials of the aircraft to be conducted at Dovecot park in Forgandenny, a thirty acre open space.

Milne gives the date of 1906 (when his exams in Edinburgh were drawing near) as the year in which Watson asked him to travel to Paris to help with the purchase and retrieval of an engine from Santos-Dumont. In the same letter he states that the French engine was delivered to Forgandenny.

Watson was studying with Professor Keunan in 1900–01 but records verify that J.B. Milne was a student at QCD from 1903–1905, after which he moved to Edinburgh to study dentistry, so presumably the exams referred to in 1906 in Edinburgh would be his first year exams.

According to the son of John Doe, the engine was delivered in a wooden packing case to Errol, not Forgandenny. Lawson Doe remembers the packing case well as his father used it as a toolbox for many years.

It measured 110 x 60 x 60 cms (43 x 24 x 24 inches). The French museums of flight do not have the 'double de Dion' engine so it is not possible to measure it accurately, but photographs show it as a tall engine, approximately one metre high and about 40 cms broad (39 x 16 inches). The fate of that engine is thus not accounted for.

W.R. Gibbs [31] and J.Y. both name the engine as a Dutheil-Chalmers.

The initiator of this company was Louis Dutheil, a mechanical engineer who began building motor engines and motor cars in 1895. In 1897, he sold the patents to an English syndicate. In 1898 he began to build light engines, mainly for motorcycles, and several of these were used in experiments with dirigible balloons and early attempts at aircraft [41] (but further dates and details unspecified).

In 1902 Dutheil entered into partnership with R.A. Chalmers, an Englishman educated at Harrow who by that time had "spent many years" in studying the complex problems of aeronautics by means of practical experiment. The Cie Dutheil et Chalmers was registered in 1903.

It seems that the firm was prominent and well-respected in its field of internal combustion engine design and manufacture. These ranged from 8 to 200 hp with a reputation for reliability and durability; but they also produced transportable engines – in which power/ weight ratio was crucial. These engines were supplied to the French army for sapper activities and power supplies for wireless telegraphy as well as irrigation, agriculture, private electricity supplies and use in small marine craft.

The advent of R.A. Chalmers in 1902 resulted in the firm's interest in aeronautical power units and it had produced already an 8hp, a 12hp and a 60hp engine with this in mind by the time Santos-Dumont prompted the 18hp model. Santos-Dumont modified the design and fitted the engine to his No19 (the monoplane which flew at St Cyr in 1907 and was the prototype for the Demoiselle).

The company incorporated Dr M.B. Boyd, an expert in mechanico-chemical engineering who had applied his expertise with reference to aviation since 1897, and became Dutheil, Chalmers & Boyd Cie. around 1908. About this time they introduced a series of engines designed primarily for aeronautical application (Moteurs Speciaux pour l'Aviation). These comprised 1, 2 or 3 twin cylinder, horizontally opposed units which could be coupled together, giving 2, 4 and 6 cylinder water-cooled engines of 20, 40 or 60 hp respectively. The 2 cylinder model was priced at 4,500 francs, complete and ready to start.

Descriptions in the sales material suggested that the horizontally opposed feature was a new design, but although the Company had doubtless refined it to minimise vibration and maximise the power/ weight ratio, several makes of car had featured the layout even prior to 1900 and Duthiel Chalmers' earlier 8hp unit was an air-cooled horizontally-opposed engine, which weighed only 25lbs (11.4kg) complete [42].

Records show that Santos-Dumont purchased two 8hp units and three 18hp units. Presumably the 18hp units were destined for his monoplanes 19 and 20 and probably as substrates for a later modification for the 'Demoiselle' engine which he had manufactured in the Darracq factory. Darracq subsequently claimed the design as their own which sparked off a legal action. Santos-Dumont was not interested in royalties, but he was interested in loyalties!

The purchase dates are not recorded.

The 8hp units were earlier engines and appear from photographs to have fewer cooling fins than the 18 hp engines.

Santos-Dumont's airship No7 had two propellers, both driven from a single 60hp Clement engine, but his only twin-engined construction as recorded by Wykeham [43] was No11 which was designed initially as a glider and subsequently as a heavier-than-air monoplane with twin engines.

It never flew in either form and plans did not survive, but it was apparently the subject of unsatisfactory trials as an engine-less float-glider towed behind a motor boat. The planning stages were in 1904. Hoffman does not deal in any detail with the engines that Santos-Dumont used in his airships, but in relation to No11 he states:

> His sketches showed that he planned to add two engines to it to turn it into a twin-propellor aircraft, but he never did this because of the stability problem.

By later on in 1905, Santos-Dumont had already concerned himself with a helicopter design; the construction and (engine-less) testing of No13 airship; and the construction and test flying of No14 which had a Clement 3.5hp engine.

It is tempting to believe that, if engines were purchased at all with a view to fitting them to No11, these might have been 8hp Dutheil Chalmers units and that he might have sold the pair to Watson, but this would not fit with the fact that the original engine was referred to as single by Kerr Sturrock and J.Y.; that it was delivered to Errol in a single packing case sufficient to accommodate only one engine; and that in 1905 it would not fit the description 'old balloon engine' (singular) or Kerr Sturrock's description of a second engine being added at a different time to the first.

Kerr's evidence is likely to be reliable about there being a French engine fitted as a single unit before that, and he was, after all, the man who was fitting the propellers to it. The photographs of plane No1 in 1906 leave little doubt that it was eventually fitted with two Dutheil Chalmers units and his remark about the considerable re-arrangement of engine(s) and propellor rings true and indicates that these modifications must have taken place around 1905.

Photographs from the Musee de l'Air of Santos Dumont's No16 airship show construction with what are clearly two Dutheil Chalmers horizontally opposed air-cooled units. This machine was a composite airship and aeroplane with a gas bag *and* wings and tail. According to Wykeham [41] it was fitted with a 50hp Antoinette engine. However, the Musee de l'Air states that there were two versions, as substantiated by the photographs.

Number 16 was destroyed on the ground in 1907 before it ever flew. Again, it raises the possibility that the engines were those purchased by Watson for Plane No1 but this does not correspond with the datings so far indicated. It does, however, establish the point (*vi*) that such engines were already in production by 1907.

It is not known whether the patents sold by Dutheil in 1897 to the British syndicate were manufactured under a different name or even whether they were horizontally opposed units. These were relating to engines designed before Chalmers joined Dutheil and may not even have been horizontally opposed twins. But if they were, it is possible that Watson may have purchased one in Britain to begin with, and subsequently purchased another from Santos – Dumont (this seems unlikely as one would imagine that to couple two units together, the two would have to be identical or very closely similar. Certainly later Dutheil-Chalmers engines were manufactured in twin units which could be coupled together as pairs or threesomes. In any case, the two engines used by Watson were not coupled as a single 4-cyl. bank, but were placed on either side of the propeller. But, as to a first engine, none of the Dutheil-Chalmers engines up to that time would comfortably fit into the dual description of "little French engine" or "old balloon engine", as Santos-Dumont had not previously used Dutheil-Chalmers engines.

Then there is the mystery of why J.Y. chose to delete the connection to Santos-Dumont in all the later journalistic references.

Plane No1, pictured at Forgandenny (J.Y. dates the photograph at 1906) can be seen to have what appears to be two horizontally opposed twin units, coupled to a propellor between them. This corresponds closely to the drawing of the engine by John Gourlay.

Kerr B. Sturrock, who made and fitted the propellers states clearly that plane No1 was fitted with a 10hp French engine which was insufficiently powerful. (It was powerful enough to smash several of the earliest propellers!) "It was then that another 10hp engine was coupled up with the existing one. This alteration made a complete change of propellers and caused quite a lot of alterations".

Sturrock does not himself state in correspondence that the first engine was a de Dion, but said so to J.Y. in a conversation recorded by J.Y. in which Sturrock referred to making the propellers for a de Dion engine, with details of the flange bolts used. He described there being five such bolts. The Dutheil-Chalmers engines apparently had six.

As previously noted, Sturrock states that the work was going on "long before" 1905, and "it was off the ground but would not stay".

This leaves the question of whether there was an earlier French engine which was replaced by two Dutheil Chalmers units, or was there a single Dutheil unit to start with, to which a second was later added?

Two other possibilities are that the engine was French, but did not come from Santos-Dumont; or that it was an engine that Santos possessed but did not use in an airship and sold it on to Watson. J.Y., after the initial preparation of his dossier, used an ink pen to alter or delete some of the dates and other details. One consistent alteration is the scoring out of statements to the effect that an engine was purchased from Santos-Dumont, although at one point he obviously thought that it had been. The relevant wording has been scored out in press cuttings from *The Daily Record* (26 February 1949 and 3 March 1954); *The Manchester Guardian* (15 December 1953); *Dundee Evening Telegraph* (15 December 1953); *Scottish Daily Mail* (16 December 1953); and in J.D. Leslie's article in *The Scots Magazine* (October 1953).

A definitive answer is unlikely to be forthcoming now, but the best fit would seem to be that there was an engineless rocking wing glider in 1903, to which a French engine was added and that subsequently two Dutheil Chalmers 8hp units were substituted, probably around 1906, as they were certainly in place in the Forgandenny photograph dated 1906. Comparing the propellers in the photograph with those on the twin engined unit designed by Santos-Dumont for his airship No16, it seems likely that those were the same engines. Photographs of No16 emerging for launching in 1907 show a single propeller, presumably attached to the 50hp Antoinette engine, suggested that the other version was the earlier one and the engines may have been sold to Watson.

PLANE No2

Plane No2 was also at Errol and at Forgandenny. The possibility is that the long flat area of ground without the threat of surrounding trees that existed close to Forgandenny Station – known as 'The Haughs' – was more suitable for making the take-off runs. Bell-Milne's letter to J.Y. suggests that the arrangement to use Forgandenny was for testing Plane No1. The ground that his uncle made available was Dovecot Park in the estate grounds and it seems more likely that testing would take place there first and the move to the Haughs would be later and would involve plane No2 which did not involve the catapult launch. However, Gourlay states that the Haughs site was used before Dovecot Park [44].

The following article appeared in the *Dundee Advertiser* [45] in 1909:

DUNDEE AEROPLANE

Keen interest at Forgandenny

Trial Trip this Week

The inhabitants of the quiet little village of Forgandenny, Perthshire, are agog with excitement at the prospect of witnessing the trial flight of an aeroplane in their midst. The appearance at the railway station on Thursday of an apparatus suggestive of a flying machine attracted unusual attention and when it became (rumoured) abroad that the machinery in question was none other than parts of an aeroplane, the interest increased tenfold. It was soon learned that the owner of the aeroplane, Mr Preston A. Watson, of the well-known firm of Messrs Watson & Philip, had selected as the scene of the operations a field in close proximity to the railway station and from then till now the villagers have been kept in a ferment of excitement at the prospect of witnessing a trial flight of the machine. Mr Watson has already made active preparations for a trial of his machine and the platform has been laid down. The engine, etc. are stored in a tent which is pitched on the banks of the Earn. In the evening it was learned that two suspicious individuals who, it was thought, would not have been averse to appropriating or damaging the contents of the tent were in the vicinity and Mr Watson deemed it advisable to keep a strict watch on his aeroplane which he guarded all night. Our Perth representative had conversation with Mr Watson early on Saturday morning but the young aviator declined to furnish any particulars of his proposed flight. It is learned, however, that a trial will be made towards the end of this week* should conditions prove favourable.

* The weather in the ensuing week seems to have been very bad – so bad that on the Thursday severe gales damaged shipping. The Blackpool Air Rally was held at that time and was reported in the subsequent editions of the *Dundee Advertiser*. These reports indicated blustery weather in the following two weeks or so in the Blackpool area but this did not prevent a good deal of flying. There were no subsequent reports from Forgandenny, however

This report provides considerable difficulties in fixing the timing of events, as it calls into question a number of the eye-witness accounts. It reads as though the inhabitants of Forgandenny had never witnessed the like, but the fact that plane No1 had undergone trials at Forgandenny from 1905/6 onwards is well corroborated. Operations were first moved to Dovecot Park to preserve secrecy. It may be that the security was extremely tight, but it seems unlikely that in a small community like that, no word would have got around regarding such strange 'goings-on'. Certainly the Dovecot Park trials would have been out of the public eye, whereas an experimental flying machine on the Haughs would have occasioned quite a public spectacle and would not have been in the least secret.

It seems very likely that the aeroplane in question was plane No2. The patents were lodged on 1 January 1909. Most of the information from those who were involved first-hand suggests that plane No1 was in the hands of the Dundee Aero Club by then and that plane No2 was the focus of attention. However, the harder evidence – e.g. the availability of the Humber engine – suggests that plane No2 could not have been flying as a powered machine before late 1909 at the earliest. The press article refers to an engine, suggesting that it was not a trial of an engineless plane No2. But the reference is a collective one about the equipment stored in the tent (The engine, etc…) suggesting that the reporter had not actually seen an engine and it would appear that Watson himself was very tight-lipped about the whole affair. Humber probably had the 3-cylinder engine by October, but unless Watson had been in contact with them before it came on general sale, it is difficult to see how this could have been the engine in the report.

If this were the very first public appearance of plane No2, then the claim that plane No2 was demonstrated at Errol Games in July 1909 must be inaccurate either as to the year or as to which plane was on show. It could have been Plane No1 under reconstruction, but what was described was that the engine was run from time to time during the show and plane No1 was engine-less by this time. W.R. Gibbs [31] claims that the Aero Club had an engine for plane No1, but there was never a suggestion that they ran it.

There is no doubt that plane No2 was back at Errol and a good deal of the development was done there. It may therefore have been 1910 that it was at the games and it is possible that because patents had been lodged by then, the need for secrecy was removed. As far as military secrecy was concerned, in 1909 the British military establishment had advised the Government that these flying contraptions were of no strategic significance (this advice was later withdrawn amid some embarrassment!).

How one interprets the significance of this press article is crucial. If the idea that this was the start of operations at Forgandenny and the plane in question is plane No1, then it gives strong support to Gibbs-Smith's assertion that the whole story has been unconsciously pre-dated. It also means that all the eye-witness accounts, including those corroborated by such things as dates of marriage and relation to other well-marked events, e.g. Kerr Sturrock's description of adding the second engine and the timing of restoration of plane No1 by the Aero Club, would have to be dismissed as fiction – not so easy to do as they were independent of one another and consistently placed the dates earlier.

There is no information that allows accurate fixing of when plane No2 was started but there are a number of accounts from individuals who were involved with it, suggesting that No1 plane was still 'in service' until 1906; but if so, it must have been put into storage at Errol about that time, as it had been abandoned for some time before the restoration in 1908.

The second plane was piloted by several accomplices and J.Y. explains this on the basis that Watson had promised his wife not to fly the second machine and, although not related to a specified machine, he mentions a promise to his father along the same lines. That might have been later and applied to plane No3.

James Manson, foreman at Watson and Philip before World War I, was identified by the management there as being very versatile at practical tasks. He was recruited by Watson to effect repairs on plane No2 during the test flights which, according to Manson, was built in 1908.

Manson was later more deeply involved in the building of the third aircraft.

John Logie, as a boy of twelve and thirteen helped at Errol in 1907–08.

The plane with which he was involved was definitely flying and he mentions the wheels – indicating that it was plane No2 and implying that this plane was also flown at Errol. Some of the photographs also indicate that this was so. Logie describes one incident when the plane was flying at about sixteen feet and climbing, but just failed to clear some trees when Watson was negotiating a gap through them – and the plane crashed.

J.Y. also dates plane No2 as 1908.

John Doe occupied Muirhouses farm house, Errol in May 1910. By that time the shed which housed plane No2 was already in use. His daughters were young at the time and were given sweets by Watson when they came to watch him at work. The childrens' nurse, Lizzie Law, brought him cups of tea and was sometimes recruited to stitch torn fabric.

David Barclay assisted with plane No2 and remembered the engine being stripped to send the pistons and connecting rods to Dundee for lightening. He stated that the plane was initially kept in a shed at Muirhouses but was later moved to the corner of a field on Leys farm, presumably because this was more conveniently situated for the trial site.

Barclay left Errol on 22 January 1913 to work for the Caledonian Railway. He eventually became Inspector of Bridges for the railway.

John Harris was an apprentice gardener at Rossie House (Forgandenny) from 1907–1911. In 1908 he remembers helping to carry the plane from shed to launching ground. It is not stated that this was plane No2, but the dates suggest that it would be, indicating that it flew both at Errol and at Forgandenny.

Plane No2 was of similar, but not identical design to plane No1. As it seems clear that plane No2 was under construction or flying in 1908, i.e. by the time the restoration on Plane No1 began, it is also clear that it was not constructed by adapting or cannibalizing plane No1.

Watson retained the tower structure with a similar fuselage. Like No1 the fuselage sloped up from horizontal to hold the box-kite tail section well above the ground when the aircraft was parked. The triangular patterns of bamboo were retained with some increase in the small bracing pieces and a reduction of the long bamboos to four lengths.

The skids seem to have been retained but were now shorter, lighter and held off the ground by four wheels, one at each corner of the base of the tower.

The control system was retained both in respect of the control surfaces and in respect of the stick and cabling. Like No1 there was no vertical rudder.

The wings were much the same as No1 except that the covering was continuous across the fuselage, giving the impression of a single lower wing. The camber of the upper surface seems to have been greater. The pivoting upper wing appears to have been much the same as before.

There was none of the fuselage projecting forward of the leading edge of the wing. The fuel tank was mounted at the top of the tower under the top wing. The impression given was of a somewhat lighter construction achieved by more efficient bracing.

The engine was a Humber 3 cylinder semi-radial (or 'fan') layout, said to be rated at 30hp, and was mounted as a tractor unit between the two forward uprights of the tower with the propellor flange projecting.

In a way, the design similarities between the two planes vindicated the design of plane No1. The main change was the use of a more powerful engine and, as Sturrock put it, the second plane "was well in the air".

Once again, the details of the engine could help in tying down the chronology; and once again this aspect throws up some difficulties.

In 1868 Thomas Humber founded a small company in Sheffield to manufacture bicycles [46]. (It may seem odd how many connections there are between cyclists and cycle-makers on the one hand and pioneer aviators on the other). The Brothers Wright, Samuel Franklin Cody, Glen Curtiss are some examples. Perhaps it had something to do with – in the case of manufacturers – the desirability of lightweight construction; and in – the case of cyclists – the ability to balance using subtle and co-ordinated shifts of body weight).

Thomas Humber was a gifted blacksmith whose reputation spread locally by word of mouth and who had no great ambition to expand into 'big business'. But he eventually expanded, moved to Hull and thence to

Nottingham. There the company progressed to motorcycle construction and thence to a 4-wheeled "Voiturette". The Company then moved on to car manufacture and were eventually to become one of the largest British producers with a reputation for quality, reliability and comfort. Behind the public face of respectability and stability were some paradoxes. Despite some quite severe boardroom scandals in the early days, their public image was a conservative one, but nevertheless they embarked upon some developments which were not so soundly based and turned out to be relatively short-lived.

One such was the establishment of an aero division, based in Coventry. The matter was reported in *The Aero* of September 1909 [47] stating that fifty monoplanes of the Bleriot 'Cross-Channel' type were to be produced costing £400 apiece. By December 1909, Humber was able to ship two such aircraft to India.

Humber offered various engines for sale as separate items. One of these was a 30hp semi-radial, or fan-type, air-cooled engine priced at £125. The cylinders were arranged at 60 deg. to one another. This engine fits the eye-witness descriptions and the photographs of the engine in plane No2.

Humber had previously produced a 9 hp 3-cylinder engine in 1903, but this was apparently not a success and was abandoned after a short time. It is not known exactly when Humber began production of the 30 hp 3-cylinder engine, or when they first offered it for sale. It is almost certain that it was an identical engine to the one designed by Anzani that had been in production previously, and was being manufactured under licence (*vi*) Bleriot used one on his cross-Channel flight in 1909. According to Gibbs-Smith, this engine did not go into production until 1909, but de Pischof's plane of 1907/8 (which was not a success) was fitted with a 3cyl engine of this type rated then as 25hp and Bleriot fitted one in the plane he used to cross the channel in 1909. Thus it is difficult to be sure how early this engine bearing a Humber marque would have been available to Watson, but it could not have been much earlier than 1908/9. However, it undermines the statements about plane No2 being airborne in 1908.

Alessandro Anzani (1877–1956) was a gifted, self-taught engineer from

North Italy. He moved to France in 1900 where he lived most of his life, developed his skills and began his own business. His engines earned a high reputation for reliability and value for money, which arose in large part from motor-cycle racing successes. He founded the business of The Anzani Company in Paris in 1907 and he – and it – prospered [48].

There seems little doubt as to the merits of his engines (there were numerous types) but his rise to fame was greatly boosted by his association with Bleriot and his part in the Cross-Channel flight in July 1909, encouraging an uncharacteristically hesitant Bleriot to set out on the day. Although the engine overheated on the way, a providential rain shower helped to cool it and the Anzani reputation for performance and reliability was not merely preserved, but given a tremendous boost.

The main features of the Anzani engines were that the majority were air-cooled allowing lighter weight; and they had a reputation for reliability without being outrageously expensive. Anzani originally designed them to fit on bicycle frames before they were used to great effect in motor-cycle racing. These were mainly 2-cylinder engines but he added a third cylinder to create the more powerful 30hp engine that Bleriot used.

This proved an extremely popular engine and the Anzani Cie. issued a licence to Humber to manufacture it under their own name. This must have been issued by the original Anzani company, based in France because British Anzani were not formed until November 1912. British Anzani finally went into liquidation in 1980.

PLANE No3

What little is known about the start of the third aircraft comes from an article in *The People's Journal* of 7 May 1960, based on interviews with James Elder Manson when he was seventy-two years old. As already mentioned, Manson helped with maintenance and repairs on the second plane in the period of time following its completion (put by Manson at 1908). It appears that Manson had proved himself so useful in that capacity that Watson managed to have Watson & Philip release him to work full-time on the third aircraft.

Manson was obviously very able and after the outbreak of war, he enlisted, first of all in the Black Watch, then changed to The Royal Flying Corps and finished up in charge of engine fitters in Peterborough.

According to his account, plane No3 was embarked upon with the object of showing it "at an exhibition in Paris at the beginning of 1914, where a prize was to be awarded for a safety device against side-slip in the air".

Watson abandoned the use of bamboo and used instead tubular aluminium for the fuselage and for the tower structure. There were only two long members, in the same horizontal plane, running from the engine bearers at the front to a horizontal tailplane arranged to act as an elevator. These 'longerons' ran horizontally back and formed the fuselage. They were parallel to the ground when the plane was not flying and were braced by wires from the top of the tower structure to points immediately forward of the tailplane.

The 'tower' structure was also altered somewhat. The aluminium tubing was arranged as two A-frames, one forward of the lower wing and the other immediately aft. These were connected at the top by the structure that pivoted the shorter top wing. The base of the A-frames was at the level of the lower wing and from the lower extremity of each leg the tubing was continued perpendicularly down to support the plane on the four wheels attached to the bottom of the perpendiculars. This basic shape was suitably braced for structural strength. It did, however, alter the shape of the aircraft, although the basic layout of lower wing, shorter pivoted upper wing, forward-placed engine and tractor propellor were retained.

Watson himself gives a full description of this plane, accompanied by photographs, in his article in *Flight* magazine of 15 May 1914 [49]. The article (*see appendix*) relates primarily to his rocking wing device, but provides the clearest picture of the construction of the aircraft. The photographs are probably taken in Gourlay's Foundry before the plane went to France.

The control system was also retained for the most part. The control stick was attached to the rocking upper wing and also to an elevator on the tailplane. The box kite structure used in planes No1 and No2 was replaced by a monoplane tail. There was no vertically pivoted rudder. Some photographs indicate that Watson may have modified the original tailplane of plane No2 to this configuration before plane No3 was started.

The engine used, according to J.Y., was a 60hp 6-cyl air-cooled radial Anzani. There are two main possibilities. By this time British Anzani had been formed by the parent company in France in November 1912. The premises were in Willesden which began to see a concentration of aeronautical manufacturers prior to and during the First World War. They continued to produce a 3 cylinder 30hp engine, but this had cylinders at 120 degrees (and is clearly not the engine used in plane No2).

British Anzani produced a 50–60 hp 6-cyl. Radial priced at £372 and this best fits the reported facts. They also made a 40–45 hp 6-cyl radial priced at £300. The 60hp engine would certainly not merit the description 'underpowered'; and neither, indeed, would the 40–45 hp. The photographs show a radial engine and there is no particular reason to doubt J.Y. on this point.

CHAPTER V

The Patents

Preston Watson had two patent applications accepted. The first of these, patent no. GB190723553, was applied for on 25 October 1907 and was granted on 8 October 1908. It related to an idea for raising and propelling a heavier-than-air machine through the air.

Watson proposed a system of paddle wheels, the blades of which altered pitch by feathering on the upstroke and assuming 'coarse pitch' on the down stroke in order to raise the machine, or different combinations of these to propel it forwards. The alteration was under control of the operator.

This now seems far-fetched, or even ridiculous, in the light of the systems used in modern aircraft for propulsion and steering control, but it carries the principles on which are based the Voith-Schneider marine propellor and more recently the fan wing aircraft (see Chapter VI).

The only record of this is the patent itself (see Appendix I). It is not mentioned in any of the surviving correspondence, or by J.Y. in his later documentation.

The second one, patent no.GB190900047, was applied for on 24 July 1909 and was granted on 16 December 1909. This one related to the control system devised by Watson to provide lateral and attitude control by means of a single control lever. The lateral control was provided by means of the rocking top wing which he incorporated in plane number one. The attitude, or elevator, control was provided by a pivoted tailplane, initially of box-kite construction but in plane number three, by a single plane surface. This was controlled by rods or wires from the 'stick' control.

The illustration in the patent document (see Appendix II) is clearly plane number one, identifiable by the shape of the 'fuselage'; the box-kite tail; by the arrangement of the engine with the propellor placed between the two halves of the engine; and by the absence of wheels. Skids are shown in Fig.2 of the patent diagrams but under-emphasized on the other illustrations – presumably because the patent applied for was not for the aircraft itself, but for the rocking wing device. It is clear from page 1, lines 7–9 and page 2, lines 12–14 of the Complete Specification that the aircraft is not the object of the patent. The words "plane" and "main aeroplane" refer to what now would be referred to as a wing.

Watson uses the word "rudder" for what is now called an elevator (page 2, line 26), and the term "vertical rudder" for what is now termed a rudder (page 3, line 18).

If previous indications of the chronology of events is correct – and there is no arguing over the validity of the dates on the patent – how do these fit together?

It would appear that there were gaps in Watson's aircraft building activities, perhaps attributable to other pressures, e.g. his marriage in 1905, family demands, his incorporation as a partner in the family firm, and the availability of financial support from his father. At some point his father had made him promise not to fly the machines.

Indications are that plane No1 was out of service by 1906 and plane No2 was in the planning and construction stage, if not actually flying, in 1908. Perhaps he illustrated his patent application using the machine that he already had had off the ground which had enabled him to test the system to some extent. Perhaps he had simply drawn a 'generic' flying machine to illustrate his invention.

The patent agent was George C. Douglas & Co., consulting engineers and chartered patent agents, of 41, Reform Street, Dundee. The firm was established in 1887.

George Douglas was born and educated in Inverness. His engineering career started with the Highland railway and later with the North British Company in Cowlairs. From there he became a draughtsman, progressing to be chief draughtsman in a Glasgow firm of engineers. While in Glasgow he attended the College of Science and Art and the Technical College, where he achieved distinction.

In June 1883 he moved to Dundee as assistant in Thomson, Son & Co. of Douglas Foundry, setting up his own company of consulting engineers in 1887. He held a position as lecturer in mathematics and engineering at the YMCA Science School in Dundee. With Sir John Leng he was instrumental in encouraging the development of new industry and in improving the patent laws [50].

The preparation of a patent application seems not to have changed greatly in a century, although the available tools nowadays – digital photography, computer imaging and editing, etc. – render it easier and faster. Discussions between inventor and agent constitute the initial process which need not involve a lengthy process in the case of a simpler invention. Most of the applications by larger businesses are brought promptly to the agent and deadlines are set for lodging the application. Timing can be much more erratic in the case of private individuals and it is not unusual for inventors to 'incubate' their ideas for considerable lengths of time.

In 1908/9 the drafting of the application involved preparing the material for the commercial printer. As this was relatively costly, proof reading and diagram drawing and checking would be painstaking and time-consuming.

There is now no way of knowing when Watson first approached George C. Douglas & Co., but it is likely to have been some months before the date of the first submission (Provisional Specification) dated 30 December 1908.

CHAPTER VI

The Claim

James Yeaman Watson, Preston Watson's brother, assembled in the early 1950s as much evidence as he could – much of it from recall from eye witnesses and those involved in building, testing and flying the Watson planes – in an attempt to tie down the dates and nature of the early flights. His initial idea was to press a claim that his brother had flown before the Wrights.

After assembling this and presenting it, he met with counter-arguments and (it must be suspected) a degree of understandable resistance from the establishment (*vi*) and it may have been this that persuaded him later to alter some of the dates in the original material. This, of course, makes analysis much more difficult, but at least he indicated honestly the areas where he was in doubt and what emerges in the correspondence is that he is aware of his own liability to prejudice and his need to allow for this and to be intellectually honest.

J.Y. admits that he himself was not much involved with the early design and development of plane No1 as he was absent in the USA for some of the time; and therefore some of his own recollections are incomplete. Nevertheless, he was involved in many of the flights, including some of the early ones. He piloted some of these early flights as he was lighter in weight than his brother.

It would appear that he began the process in 1948 with a letter to the Science Museum dated 21 December 1948 because there is a reply on behalf of M.J.B. Davies, dated 15 January 1949, thanking him for photographs and other material referring to the earlier letter. This is followed by another mentioning the engine as being rated as 10hp and acquired in 1902 (corrected in pen to 1903)

Whether this could have been picked up by the press or (more likely) that J.Y. had written a newspaper article by this time is not clear but a column entitled J.D. Leslie's Log was published in the *Daily Record* of 26 February 1949 on the subject of 1903 flights by Watson. This article contains a number of inaccuracies but it is possible that these came from J.Y. himself as at this stage he had not yet sifted all the sources of information. J.D. Leslie wrote again in the *Scots Magazine* in 1953 and some of the inaccuracies were persisting [51].

Naturally the claim generated some public interest and there followed newspaper items, press correspondence and magazine articles. Most numerous were those published in the *Dundee Courier* and *Advertiser*. Some of these were letters from individuals (or their relatives) who had had first – or second – hand involvement in these early activities, but many were re-runs of previously published material. Mistakes and inaccuracies crept in, presumably because of some inconsistencies between some of the recollections.

If these articles are reviewed serially over the whole period after 1949 it is clear that some of the inaccuracies eventually take on a mantle of truth, perpetuated in subsequent writings and internet sources.

For instance, the series by J.D. Leslie contains a number of inaccuracies which have persisted into various subsequent press articles. Although articles in *Meccano Magazine* (1957) [**38**] and *Leopard Magazine* [**52**] do give a good general overview and are more cautious in drawing conclusions, there are some facts which turn out to be either untrue or at least debatable. The more recent article in *The Scots Magazine* by James Allen [**25**] (author of *Wings over Scotland*) [**53**] which also mentions Watson at some length) is more authoritative, being based on his own researches of original sources.

It seems that J.Y. took up the task of assembling as much information relating to the first of Watson's aeroplanes as he could find. He must have let it be known that he wanted such material, judging by the surviving correspondence. Starting with his own recollections of events as a framework, he set about interviewing any surviving witnesses that came to notice. Some material was contributed by surviving relatives of witnesses who had since died. Whatever notes or diaries Watson kept at the time is unknown as none have come to light. The question of secrecy has already been mentioned in Chapter III and may have contributed to the present obscurity.

There certainly were plans (it would be strange were it otherwise) as witnessed both by David Urquhart and W.R. Gibbs, but any recordings of experimental results and performances, test–flight dates, etc., have not survived. The plans used for the 1909 patent application for the rocking wing device appear to be those of plane No1 and because they appear to be accurately drawn, they were prepared from accurate dimensions provided to the draughtsman.

He realised that he was drawing on remembered accounts of events which had taken place fifty years or so previously and tried as best he could to make cross correlations from different sources. Those relevant to plane No1 are tabulated below (see also Chapter IV). (*Figures in brackets are the age of the witness at the time of interview with J.Y., i.e. ca 1953/4*)

1   Scott Symon and John Christie (79) Both lived in Errol. Christie was ploughman at Daleally Farm nearby, between 1895 and 1903, after which he left the district. Both saw the plane flying high enough to clear the hedges and both set the date at 1903/4.

2   James Elder, a local inhabitant, was able to take J.Y. to the place where he remembers the test flights to have taken place in Errol and J.Y. was able to find the stump of the tree that had been used as a launching device.

3   William Brown (69) lived in Errol and worked on Daleally Farm 1901–3. He was able to identify the shed where the plane was kept and often saw it making hops which were high enough to clear hedges and allow him to see the underside of it. He gives the dates as around July 1903.

4   Alex Robertson (63) when he was thirteen years old (i.e. 1903), helped to move the plane from its original shed to a shed at Leys Farm. He himself never saw it flying but reports that his father saw it flying high enough to clear the hedges

5   Henry Band (75) was working at Muirhouses Farm in 1903 and saw the plane making hops. He remembers the sound of the engine (like a machine gun). He is very definite about the date, as he married in June 1903 (verified by marriage certificate) and shortly thereafter left the area. He was very emphatic on the matter and actually offered to sign an affidavit for J.Y.!

6   David Urquhart was involved with the restoration of plane No1 after it had been donated by Watson to the Dundee Model Aero Club in 1908. He remembers having the plans for it and it was always referred to as the 1903 plane. Whether the plans carried that date is not specified.

7   W.R. Gibbs was a lecturer at Dundee Technical College. He was a member of the Dundee Aero Club and had been involved in the restoration on plane No1. In a lecture in 1954, he recounted how he had actually handled the engine (which he said he could lift on his own) and he referred to the 1903 plans.

8   Kerr B. Sturrock was involved with making propellers for the first plane. He made many and experimented to achieve success. He described plane No1 as being off the ground (but would not stay) prior to his marriage in 1905, after which he was less involved.

Thus armed, J.Y. set about lodging the claim. It appears that he contacted a number of figures of prominence in the aeronautical scene of the 1950s and although some of the correspondence carries a date, it does not emerge clearly whether he had a master plan. The impression is that he did not; but rather that he cast around for a sympathetic ear with which to start off.

Oliver Stewart, Editor of Aviation, published an article in *The Tatler & Bystander* [54] covering the pioneering years of aviation. Following this, J.Y. wrote to him on 19 November 1953 about his brother's early activities. Stewart promptly replied (23 November 1953) expressing some interest and J.Y. provided him with some further detail, mentioning also that Preston Watson had been mentioned in a speech by the then Secretary to the Ministry of Civil Aviation, Lord Profumo [55].

Perhaps he was encouraged by this level of interest, and he wrote to Lord Brabazon who not only was credited with the first flight over British soil by a British citizen and held flying licence number 1, but was in 1953 the President of the Royal Aeronautical Society. The letter included details, photographs and a copy of the article in the 1953 Scots Magazine. He then travelled to London and had an interview with Lord Brabazon on the eve of the RAS annual dinner. Whether it was a pre-arranged interview or an informal approach is not clear, but it appears that the matter was received with what J.Y. took to be a cold 'brush-off'.

On the following day, the day of the dinner, he wrote to Lord Brabazon expressing his disappointment at the way he had been received, but having the good grace to wish the occasion success. The next day the following reply came to J.Y. at his London hotel:

18 December 1953

Dear Mr Watson,

I feel I owe you a deep apology, for having been, according to you, rather brusque, which I assure you I did not mean to be, but I was fussed by the approach of the great Dinner and having to make a speech and all that sort of thing.

I hope you will forgive me, as, when I said I would like to see you, it was a sincere wish.

Nor do I in any way wish to discredit your brother's activities in the early days. On this point however, we were such a close band of workers in the pioneer days, that it is very odd, to me that the name of your brother was not known. I also have had, as perhaps you know, great trouble about the early flights through A.V. Roe, who claimed many years afterwards, that he made a flight in 1908, which he never mentioned at the time although a year later he made great play about the flight he had made during that year. I suppose that sort of thing has left me rather a little bitter and the idea of another claim fifty years afterwards, really appalled me.

It was indeed generous of you to wish our Anniversary Dinner a success. I think it was.

But I am left with the feeling that I have hurt you, and that, I assure you, I did not intend to do. Please, therefore, forgive me .

Yours very sincerely,

(sgnd.) Brabazon of Tara

J.Y. comes across in this, and later, correspondence as basically polite in his correspondence and warm, 'chatty' and informal in his personal contacts.

However from a subsequent letter to Lord Brabazon, written on

Christmas Day 1953, it would appear that he must have lost his temper to some degree during a phone call to Brabazon's secretary. He wrote to apologise and say that he would take up the suggestion that he would formally contact the Aeronautical Society.

Also on Christmas Day 1953, he wrote to J.P. Derriman of the *News Chronicle* in relation to a previous interview. An article had been published in the Northern Edition earlier in December and in the *Manchester Guardian* in 15 December 1953 [56].

On 29 December Derriman replied:

Dear Mr Watson,

Thank you so much for your letter of 25 December. I now gladly send you a copy of the *News Chronicle* Northern Edition which contains my story of our interview. Incidentally, I was sorry to notice that someone in our Manchester Office had been foolish enough to write a headline suggesting you had claimed an earlier flight than the Wright brothers. In fact, you made it quite clear that this was not your claim, as you couldn't be sure of the date.

You will see that my story does make that quite clear, and I can only say that the headline was not my fault!

Naturally I shall be glad of any further news from you at any time.

Perhaps J.Y. was beginning to falter a little in front of the heavy artillery of the establishment! But a further salvo had already been fired. *The Manchester Guardian* version had come to the notice of Charles Harvard Gibbs-Smith, well established aviation historian, who replied to the editor:

Sir – With great respect to the Watson family of Blairgowrie, I would say that the visual evidence of the aeroplane photograph you published yesterday points to the machine having been built not earlier than 1906, and probably in 1908 or 1909. One might meet one or perhaps two freak anticipations in aircraft design, although I know of no twentieth-century aeroplane that does not clearly reveal its origin in its design. But the Watson machine fairly bristles with derived items.

First, Watson would have been unlikely to have obtained an engine light and powerful enough from Santos-Dumont before 1908, when the latter was using a Darracq engine in his little Demoiselle monoplane, although it is just possible Watson might have obtained Santos's cast-off 24 hp Antoinette engine of 1906, but it was almost certainly too heavy for its power output. Santos-Dumont's previous engines, which he had used in his airships, would all have been useless from the power-weight point of view.

The box-kite tail and its method of hinging to the outriggers or booms is exactly similar to Santos-Dumont's "14bis" of 1906, except that it is at the back, "a la Voisin". The boom construction itself suggests a mixture of Demoiselle and the method of Curtiss of 1909.

The undercarriage, with its curved skids lying immediately adjacent to the wheels, is so like Henri Farman's 1909 machine – the first to use this form – that it is hard to believe that the latter was not the model. The strange single rocking "aileron-wing" was probably derived from the two Curtiss-type between-wing ailerons of his 1909 machine, which flew at Rheims in that year. The propeller question is even more hopeless. It would have been virtually impossible for Watson at that time (1903) to have done the complicated work involved in making even a primitive workable airscrew.

Whatever the exact year the Watson machine was made, I do not think any historian could agree for a moment that it was constructed, let alone before 1906, and I would put 1909 as the probable date (a 9 can look very like a 3 in a document). In any case, it is the ability to make properly powered, controlled and sustained flights which forms the criterion of a successful aeroplane, and the Wrights achieved this in 1903. – Yours &c

C.H. Gibbs-Smith

Royal Aero Club, London W.1

It is interesting to compare the arguments used in this letter with those used in the later extensive correspondence with J.Y. Some of them seem rather flimsy and disappear later, e.g. he does not later re-state the suggestion that Watson copied Curtiss or Santos-Dumont.

The skies must have brightened a little for J.Y. when he received on 30 December 1953 a response from Oliver Stewart, promising to put something in Aeronautic'; and writing in the letter that he thought it was "...a scandalous thing that his (Preston Watson's) name has been so much overlooked in previous years".

An article did appear in *Aeronautic* in February 1954 [57] which mentioned Watson and some of the material collected about his pioneer activities. Gibbs-Smith, in correspondence, claimed the authorship. The article must have attracted the attention of J.W.R. Taylor, who then contacted J.Y.

John W.R. Taylor [1] was a prominent member of the aeronautical establishment. He had been a member of the design team at Hawker Aircraft, Ltd. and became chief compiler of *Jane's All the World's Aircraft*. He was author of over 200 books on aviation and space travel and was a Fellow of the Royal Aeronautical Society, Royal Historical Society and a member of the Society of Licensed Aircraft Engineers and Technologists.

In the course of the correspondence he expressed interest. He thought there was insufficient material for a book, but indicated that he would make mention of the facts in a forthcoming 700 photograph 35,000 word *History of Flying* to be published by Hulton Press. In his letter he stated that "Nothing gives me greater pleasure than to be able to right a serious omission in aviation history".

There were two titles subsequently published in which Taylor had a leading role as author or editor *History of Aviation* (1972) and *The Story of Flight* (1970) [1]; but Watson was not mentioned in either.

Another contacted was Don Dwiggins, aviation editor of the *Los Angeles Daily News* at the time. He had been a flight instructor in the RAF in World War II. J.Y. was in correspondence with him in 1954 but it is not clear who made the initial approach. Dwiggins requested material on Preston Watson as he was preparing a historical work. Dwiggins was a prolific writer and he may have given mention to Watson in a journalistic forum. It is also possible that J.Y. did not send him the material after all, as there seems to have been no further correspondence.

J.Y. refers in a letter to the fact that a collection of relevant photographs had been given to a London photographer (Maurice Allward) for processing. It is not clear what was actually intended, but presumably it was to increase the impact of the claim and be prepared for a serious enquiry into it.

The photographs do not seem to have survived as any sort of collection and most of the surviving photographs are of snapshot quality. This is a great pity because it is likely that Allward would have done his informed best to bring out the technical features in the material supplied. He was not simply a photographer – it seems he was a valued authority in the aviation field and had been invited by *Flight Magazine* to review Sir Philip Joubert's broadcast series on War in the Air. (In a subsequent letter to J.Y. in 1955, Allward refers to the fact that J.Y. had had professional help in organising the data. It is possible that it was Oliver Stewart who helped him in this respect).

Oliver Stewart was certainly of great support to J.Y. In 1954 he wrote to him to say "…and I can assure you that we shall strongly support your valuable campaign for full recognition of your brother at all times". Later, in September 1955, in response to receipt of a copy of J.Y.'s summing up of the evidence, he wrote once again a very supportive and complimentary letter.

It was evident that J.Y. had spent a great deal of time and endeavour on the whole affair and had met with a good deal of resistance and discouragement. In his reply to Oliver Stewart he thanks him for support which "…kept me going when the information I dug up from the past got difficult to sort out".

Stewart had published in the January, 1954 edition of the *Tatler & Bystander*, an article which dealt, *inter alia*, with J.Y.'s activities. He sympathetically described J.Y.'s approach as reserved, reasonable and courteous and although he refrained from expressing his own opinion, he did not dismiss the Watson achievements [58].

The matter was taken up in far-off parts of the globe. As well as Dwiggins in Los Angeles, articles appeared in:

The Bulawayo Chronicle, 16 December 1953

The Daily Colonist (Victoria, BC, Canada), 17 January 1954

The Sydney Sun, 12 January 1954

On 16 November 1955, J.Y. gave the address at the Annual General Meeting of the Strathtay Aero Club at Scone Aerodrome. He titled it, "Final Summing Up of the Efforts of Preston Watson to become Airborne on a Heavier Than Air Machine" It was scripted as follows:

> It was in 1898/9 that while the Newport Rugby Football Club team were practicing running and sprinting on the Dundee Esplanade, just where the Tay Bridge enters the water, that Preston could be found seated on the parapet wall watching the sea gulls hovering overhead. When some of the team said to him, 'Come on, Pres., and train instead of sitting there' the reply they got was 'Someday we will fly like these gulls'. Little groups of the team gathered together, and pointing their fingers to their heads, remarked, 'Pres. is going dotty'.
>
> Preston was one of the best of our fifteen and a powerful player – the team photograph at the beginning of this Album* is to give those interested an idea of his height and build.
>
> Some years later much of Preston's spare time was taken up shooting gulls.
>
> The wings of these birds he glued in various positions, loaded their heads with drops of lead and could be seen in the summer evenings dropping these small gull models over the road bridge which crosses the railway line at the far easterly** end of the Esplanade.
>
> He was clever at school and when he finished his High School training he tried to get further information by attending classes at the University College, Dundee. There he studied higher physics making many calculations about wind pressures.
>
> I remember a discussion with Preston about the wind drag on two mail

* This was presumably the album eventually donated to the McManus Galleries in Dundee and would in all probability have been circulated to the members assembled for the Club meeting.

** Watson used the railway bridge at Ninewells which is at the West end of the esplanade, but may well have used the bridge over the line to Perth at the East end also. Both would be within reasonable walking distance of his home, the bridge at the east end being near the house in Roseangle and the Ninewells bridge being near "Balgowan" where his parents lived.

railway trains passing each other at high speed. These calculations were made of the algebraic type and far beyond my comprehension.

In this album a great deal of time and effort has been spent in an honest endeavour to pin down the exact year and month the earliest airborne flight by a human being took place. There is a true saying that anyone claiming that they were the first to do anything, should be ready to leave the country in which they made the statement and hide themselves.

My final summing up is that the three aeroplanes built by Preston Watson, at his father's expense, were all of the same design. They were powered by engines fitted to the front of each of the planes – the designs being carried out after exhaustive study of the flight of birds. In the early Wright Brothers' models, in Preston's own words, these gentlemen slavishly followed warping wings and their engines were fitted so that the short flights were made by the engines pushing the planes.

There is little doubt that Preston's planes were all of a flyable type, and as far as design was concerned, well ahead of the Wright Brothers' earliest flying planes.

J.Y. had, in July of the same year (1955), written a somewhat different version. A good deal of the basic material is the same, but it was probably worded, not for a public presentation, but as what he thought was to be the completion of his mission in assembling the album referred to; and the summary version of the substance of his claim on behalf of Preston Watson (see Postscript chapter).

But the matter was to continue…

J.Y. came into direct correspondence with Charles Gibbs-Smith as a result of an article in the *Manchester Guardian* suggesting that Preston Watson had flown before the Wright Brothers. The article was prompted by the efforts of J.Y. to publicise the claim and was presumably a result of the interview with Derriman. Gibbs-Smith (1909–1981) was probably the foremost aviation historian of the 20th century. He had many publications on the subject [3], many emanating from the science museum, published by HMSO, e.g. *Sir George Cayley's Aeronautics 1796–1855*; (1962); *Aviation* (1970); *The Rebirth of European Aviation 1902–1908* (1974). His researches were very thorough and carried a great deal of credibility.

He was in direct correspondence with J.Y. by 1957. All the component letters are not available, but the following correspondence is perhaps the best way of conveying the substance and tenor of the argument and is therefore quoted verbatim. J.Y. refers to a letter from Gibbs-Smith dated 28 April 1957 which must have evoked a disappointed response from J.Y., judging by the next one from Gibbs-Smith, which does not concede any ground, but is considerate and tactful.

Royal Aero Club
119, Picadilly, W1
As from Victoria and Albert Museum
London S.W.7

11 July 1957
PERSONAL
Dear Mr Watson

I am most distressed that you feel I have a grudge against Preston. Nothing can be further from my mind: for a man who gave his life for his country, and who was an able and admirable inventor, I have nothing but admiration. But you must not begrudge me what in my work I have to preserve above all, a duty to establish the truth even when it might seem to conflict with kindness and generosity. I am a historian, and an honest historian has not only to establish facts and probabilities, but make comparative assessments. I am now nearing the end of a history of the aeroplane, and in the second part of this book I am describing all the serious claims: these claims have to be taken seriously and sifted in great detail, not only to establish the facts but to be rigorously fair to the men who are dead and gone, because they can no longer answer back.

As most of the claims involve the Wright Brothers (I will return to them below) it is just as important to be fair to them as it is to the "claimers"; they too are dead and they too have relatives. I am writing in some detail here, as I am quite sure in my own mind that you are an eminently fair person, no matter how loyal you are to your own family. (I may say in passing that I find loyalty to family in these historical subjects is sometimes a rather unhappy hindrance to investigation. It led the relatives of Montgomery into an undignified and disastrous fight with the United States Government and the whole thing was totally unnecessary and even pathetic). Also please do not mistake my careful tone for the tone of a man who has grudges and is inhuman. I will tell you straight away, there is one man in aviation against whom I have what might just be termed a "grudge", and that is Maxim; and that is because he was one of the most pompous people ever to enter our history and one who – as I describe in detail in my book – made all sorts of conceited claims for himself which every historian today knows are rubbish.

But now to Preston. Please don't forget that among the many sources, I have been reading Preston's own words in Flight Magazine (which I wonder if you have ever studied), and I think from what he says he would be the first to wish me well; he certainly would not wish to make for himself any false claims when it came to a serious published history. Do not let us quarrel over a fine man; and let us try to get at the truth; it is worth it for its own sake.

I think I am right in saying that you now claim that the machine that Preston made in 1903 or 04 was a powered machine, the first of his 'rockers' as I call them. I, on the other hand, believe that he did no such thing. I believe he did not make any "rocker" until the end of 1908 or the middle of 1909. If anyone saw or heard of a machine in 1903 or 1904 it was either a glider or a machine – not a "rocker"– with an engine which could not possibly have made it fly.

Please bear with me while I put down what I hope you will agree are basic facts:

1. Preston himself in the Flight article gives the numbers 2 and 3 to the "rockers" with the Humber and the Anzani engines respectively.

2. That leaves only the number 1, of which we have photographs showing the four-cylinder horizontally-opposed engine with the propeller revolving between the cylinders

3. Apart from the fact that the French authorities have identified the engine as a Dutheil-Chalmers of 1909, there is the important fact that the engine shown in the photograph was made especially for an aircraft, with the shaft joining the two banks of cylinders, on which the propeller revolves between them. No petrol engines were

especially built for aircraft until 1906 (the Antoinette Motors) and no engine of this kind until much later. Therefore the engine shown cannot be anything to do with the years 1903 or 1904 or 1905.

4. This engine is shown clearly in Preston's patent drawings of 1909.

5. Quite apart from the facts above (and below), is it conceivable that a man could think up an idea in 1903 or 1904, and wait until 1909 to patent it

Please do not forget that he had **already** patented an entirely different idea in 1907, quite unconnected with ordinary aircraft.

6. Preston himself says that the Wrights were "the first to fly in a practical way". Preston did not say "first to fly" or "just first to get off the ground", or any other vague phrase: he says "practical"".

7. Preston makes it perfectly clear in his own words that his "rocker" idea is in answer to the Wrights method of wing-warping. Neither Preston nor anyone else in Europe knew that the Wrights had flown in a practical way until the Wrights themselves flew in public in 1908, so he could not have thought up an answer to the Wright's system until after he knew they had flown in a practical way; i.e. until late summer of 1908.

8. In view of the above it is surely quite obvious that he thought up his answer to the Wrights in 1908 and patented it in 1909.

9. In view of Preston's own excellently written explanation of his "rockers" in the Flight article, together with his own numbered and dated photographs, there cannot be any doubt about the origin and history of Preston's "rockers". The "rocker" idea does not lose anything of its ingenuity and cleverness by being dated in this way on his own admission.

10. Since the "rocker" idea cannot have anything to do with whatever happened in 1903–4, what was it that the witnesses saw or heard? You will agree that they are very vague. They could only have witnessed one of two things; either:

a) a glider which Preston might have made, and which I myself **believe** he did; or

b) a powered machine engined with either an automobile or a motor-
cycle engine: but there was no engine at that time whose power-
weight ratio could possibly have got him off the ground and kept
him in the air; to say nothing of the enormous problem of arriving
at a proper wing shape, control system, etc.

The only good evidence of Preston ever having any engine at an early date
is Mr Milne, who, after all, was actually invited to go to Paris with Preston
to find one and – like any professional man – well remembers the reason
he could not go, i.e. exams: the date, you will remember was 1906.

As no engine he could have bought in Paris, or anywhere else, would have
been satisfactory, the aeroplane that Preston then built (or in which he
had fitted it) could have done nothing more than hop. You will remember
that even Santos-Dumont with all his wealth for buying engines could
only just drag his biplane off the ground in 1906 with a 50hp Antoinette
motor.

Then there is another important but rather painful point.

Your father at the tragic and distressing inquest was definite and precise
in saying that Preston "had taken a great interest in flying for the past
seven years". Surely if Preston had done any serious inventing, especially
of the "rocker", your father would have said something to indicate a much
earlier date. Suppose in the pain and agitation of the moment he meant
to say eight, or nine, or ten years – i.e. back to 1905 – it was still in his
mind that Preston had been really seriously concerned with aviation for
only a limited number of years until his death in 1915.

If anything important to say at the inquest in answer to the Coroner had
gone back to those early years, surely your father would have been proud
to say so.

Dear Mr Watson, please believe me when I say I try to give scrupulous
care and attention to any interesting claim; and surely most of us would
be proud enough to have any relative who was a fine man and a fine
inventor, without necessarily wanting to measure him up against the
giants of history like the Wrights. We cannot all aspire to those heights,
you know, and why should we.

And now I come, regretfully, to the point where I am afraid I run the risk of offending you, but I assure you I would hate to give any offence or pain. I notice that in your various letters, and quotations from the press, you speak about the Wrights in a way which assures me you have very little knowledge of them. The Wrights are outstanding geniuses by any standards. I don't know if you realize that their voluminous private papers have recently been published, and engineers and aerodynamicists the world over are amazed at their accomplishments. A very great deal is not known about them. If you were to realize what Herculean efforts, what brilliant work they put in over the years – quite unrivalled by any of us here in Europe – and what endless testing and trials they made, you would, I am sure, feel differently about the whole thing. The aeroplane is a highly complex vehicle and no young man of Preston's age (23) and experience in 1903 could conceivably have had any hope of achieving what the Wrights did with all their experience. This is no reflection at all on Preston. Preston later became an expert professional airman, and would, I am sure, endorse what I am saying. He was obviously a modest and charming man, and one can tell at a glance, when you read his own words, that he understood what a formidable business the first accomplishment of flight had been; and he made no claim to any such thing.

I am taking the liberty of presenting to you, with my compliments, an article I wrote for Major Oliver Stewart's Aeronautics in 1953. There are a few errors in it which I have discovered since, and which I have noted; but the bulk of it is accurate.

The Wrights started their work in 1899; They built (between 1900 and 1902) three gliders on which they made over a thousand glides, and for the design of which they made lengthy and complicated tests and calculations after studying Lillienthal, Langley, and others and finding them inaccurate. They designed their own engine and propeller in 1903; and when they flew – four times – in 17 December 1903, it was only as an intentionally preliminary test of their first powered machine. They built their second powered machine in 1904, and their third in 1905.

By the end of 1904 they could circle and manoeuvre with ease, stay up for five minutes at a time, and had made over one hundred flights. By the end of 1905 they could stay up **for over half-an-hour** at a time, and cover 24 miles; they had then a first class practical aeroplane.

When you think that Europe's best early pilot, Henri Farman, could only stay up for **one and a half minutes and only just do a full circuit in January of 1908**, you will realise what a wonderful achievement the Wrights had made. Their wing-warping led direct to the modern aileron system, and many men – including Bleriot – copied and used their warping system. You can also see why the French and English pioneers of 1908 were astounded to witness Wilbur Wright – in that same year – easily make flights over an hour, including one of **two hours and twenty minutes** without landing, a staggering achievement which, had it not been witnessed and timed at Auvours, they would not have believed. The Wrights' control and manoeuvrability was so perfect that the French aeronautical press in 1908 could scarcely find words to express their admiration, after their own pilots **Had only just made the first official wavering circular flight a few months before**, whereas the Wrights were doing circuits in 1904.

You see, perhaps, why I am very, very careful and cautious in how I approach claims in flying. If I were not, and if I were not as strict as possible, I could not claim to be an honest historian, which I hope I am, as well as being an admirer of all good pioneers, including Preston and yourself.

Please do not think too hardly of me.

With kindest regards,

Yours sincerely

(sgnd) Charles Gibbs-Smith

There is no denying the weight of authority behind this letter, and it must have given J.Y. considerable food for thought. Couched in polite and even sometimes sympathetic terms, it nevertheless not only drives home the points of the argument but it conveys the message that J.Y. is not an academic authority and Gibbs-Smith is, almost at times seeming to tell J.Y. what and how to think about it.

J.Y. replied in the following terms:

15 August 1957

Dear Mr Gibbs-Smith,

After careful thought I came to the conclusion that in replying to your long letter of 11/7/57, I could be most helpful if I **tried** to put myself in your position, viz. nearing the end of the second part of your book on the history of the Aeroplane.

After reading over and over again your many statements about the Wrights it was perfectly plain I could be of little or no assistance at all to you about these brilliant young brothers. You have sifted their evidence in great detail, and I agree with what Preston himself stated, that the Wrights were the first to fly "in a practical way". All the help I can give you is what happened at this end, and I am going to try to keep loyalty to the Watson family from leading me to make statements about Preston's efforts to become airborne that never happened.

What I have done during the last three weeks is to check the correctness of my statements made in Preston's album. I have, since I received your letter of 11/7/57, made personal calls on the following gentlemen, all of whom I was happy to find still in the land of the living:

Henry Band, Muirskeith, Cortachy: William Brown, Errol; Jas Manson, Dundee; David Urquhart, Kirriemuir, and a new witness, James Reid, Murthly Stationmaster, Near Dunkeld. The more I dug into the past, I was happy to find that my own personal statements about Preston, as stated in the Album, were proving correct.

I now propose to give you the details of my calls, and at the same time to answer your queries in your letter of 11/7/57, also statements you make in your letter of 28 April 1957.

In paragraph (1) of your letter of 28 April, 1957, you write: "In my correspondence with J. Bell Milne there is no doubt that Preston made a glider in 1903. He even describes details such as the equal span wings and the **forward elevator**". When I read such statements it proves how far wrong people who were never on the spot, can go. In the summer of 1903, there was only Preston and myself on those Sunday mornings (while Mr

Brown was having his Sunday walk) trying to find out how, what you call the rocker wing, would handle. These tests were secret. My recollection of the rocker wing is as clear as any aged man's memory can be.

The rocker wing tests of 1903, you can take it, are established facts. The details of the movements of Wm. Brown and Henry Band as detailed in this letter prove the dates and is clear evidence of what they witnessed. The engine problem and the propeller to pull the rocker wing glider is not clear in my mind, but I am sure Preston was so satisfied with his trial by falling weights in 1903 that his design of a flyable plane were up to his expectations, that he must get busy and hunt for an engine.

So far as a controversy about a front elevator is concerned, give these thoughts up. None of Preston's three planes had front elevators. This is the big difference between Preston's designs of an aeroplane compared with the Wright Brothers' machines.

Evidence of Wm. Brown, 13 Viewlands, Errol, whom I saw last week, and took the enclosed photographs. William Brown remembers what he saw on about ten different occasions – mostly Sunday mornings – was a two winged plane being catapulted with falling weights. He states there were only two persons at these trials.

William Brown left Daleally Farm in 1903 and went to Coupar Angus for six years (this is in reply to your query, where was Brown working in 1908). He came back to the Mains of Errol in 1910. In July 1911, (and this is something that Brown told me last week that was news to me) Preston's Humber engined No2 plane was parked at the gates of the "Games Field" on Errol Games Day, for demonstration purposes. The engine was running occasionally – the Preston Watson Aeroplane was no longer a secret and was open for public exhibition. Brown was married in 1912, but his fiancée was with him at the Errol Games and has clear recollections of Preston's Humber engined plane being open for inspection by the Games spectators.

This is more information regarding Henry Band's movements than is stated in Preston's Album. In 1903 Band was married to his first wife at his father's house in Chance Inn Farm, Errol. The minister was the Rev. Walter Tait. Henry helped his father with the farm work but there was too little to do. He left Errol and took up a boatman's job at the Salmon Fishing on the River Tay with Colonel Richardson. The autumn Fishing was then rented to Lord Blythwood and the Duke and Duchess of Bedford. Henry Band was boatman in 1903 to these gentlemen. Before this Salmon Fishing work of Henry Band's, he remembers standing on top of a pile of stones at the roadside going towards the blacksmith's shop at Errol, to get a better view of Preston's No1 glider being catapulted by falling weights at the Tay River side.

This actually took place before he left Errol in 1903 to go to the Salmon Fishing on Colonel Richardson's part of the River Tay.

As a historian you should have the details of FLAFF PETE beside you. Henry Band was born in the village of Collace, near Balbeggie, Perthshire, seventy-five years ago. A veteran of the South African War, he was first piper to the Scottish Horse Regiment, having piped the regiment from Edinburgh Castle to Waverley Station, where they entrained for South Africa. He returned in 1902 and made his home with his parents at Chance Inn, Chapelhill Farm, Glencarse, Perthshire. He told me he had plenty of money from his Army Pay and nothing to do. He was in the neighbourhood of Muirhouses Farm. Errol, which was farmed at that time by David Nicoll. He recalls being attracted by a noise like a machine gun and to his surprise he saw what he described as "something like locusts flying". This was actually a plane making repeated jumps resembling the actions of the locusts he had seen feeding in South Africa. Henry Band was twenty-three years old at that time. While living at Collace at the age of eight, he was very friendly with a gentleman called Peter Hay, Collace, by Balbeggie. Peter Hay was a gentleman of independent means. He made electric power by driving a dynamo, power coming from a water wheel which drove his meal mill. He succeeded in lighting his meal mill by electric light and offered to supply Collace village with electric light but the villagers were afraid. This happened in 1887, the year of Queen Victoria's Jubilee.

Peter Hay's nickname was "Flaff Pete" because he personally tried to fly like a bird. He had been known to have jumped off the end of a high wall hoping to sustain flight by some kind of flapping wings. He landed in a midden without injury. Henry Band went on to say that for hours Peter

Hay would lie on his back watching gulls flying and continue to utter "it is bound to come. It is bound to come". Henry Band's memories of Peter Hay, as he lay watching the movements of gulls in flight, brought back rather vividly to me my late brother's deliberations. Preston did the same thing in 1899. He was sitting on the wall at the edge of the River Tay at Dundee, where the bridge crosses the river, watching the gulls hovering above his head. When asked to join the rest of his Rugby Football enthusiasts in training, all he replied was, "one day we too will fly".

Peter Hay had made many small model planes powered by stretched elastic, the ordinary thread pin being wound by stretched elastic and a small propeller being attached to the pirn. Henry Band remembers running after these models, collecting them from the ground and returning them to Mr Hay. The villagers had Hay detained in Murthly Asylum, but after a short period he was released, the doctors having declared he was not insane but only another of those extraordinary types of human beings.

At the Annual Burns Supper held at Collace in January 1954, John Martin Esq., Collace, asked the company to drink "to the memory of Peter Hay, our own Collace airbourne man of the year 1887".

Peter Hay was nicknamed Flaff Pete because he tried to fly. The villagers shouted "Flaff", meaning to flap his wings like a bird. The story goes that his wife said to him "if you're going to fly, Pete, dinna fly away, just hover abune the hoose".

Henry Band remembers an incident that happened at the time of Flaff Pete. It was one night in the shoemaker's shop, a few of the villagers were gathered there. Henry Band happened to be along with his father, and Peter was on his high-horse, stamping backwards and forwards, all afire, when he stopped and said to the villagers "You scoffers of Peter will not see what is to come when flying machines will be flying round the world, and the ploughs, harrows and tools of the farm will all be going without horses" and putting his hand on Henry's head saying "Here is a generation that will see it come to pass" and when Henry Band saw Preston's flying machine flying at the Leys at Errol in 1903, he said to himself" Peter was right after all".

In paragraph 6 of your letter to me dated 26 April, 1957, the paragraph reads "The glider photographed with Preston sitting in it is the machine stripped of its engine and rebuilt as a glider in 1910". You stated that Mr Urquhart's statements were hopelessly wrong. I do hope that you will not be displeased with me, but I feel it is my duty to prove how far from the actual facts your statement is. It is not Preston who is photographed sitting in the plane – the photograph is of Mr Urquhart – it was taken at Gourlay's Foundry in Dundee, after repairs had been carried out by the Dundee Aero Club in 1908. The gift by Preston consisted of his No1. Plane with the Dutheil-Chalmers engine and propeller. This photograph with Urquhart sitting in it, in Dundee, is the original No1 Plane – it had been lying in a shed at Errol for some months. It was a crashed machine and had been put aside as finished. Preston was then busy with the building of his No2 plane fitted with the Humber three- cylinder engine.

The repairs by the Dundee Aero Club of the No1 Plane are as follows:

If you examine **carefully** the photographs which you have beside you, you will observe that Preston had to add two additional front struts to the No1 Plane. These struts were to carry the trial engine and the petrol tank, which is easily recognizable in the Forgandenny photograph. When the crashed engine was transported to Dundee to Gourlay's Yard, these front struts and the engine were removed and the photograph of the reconditioned plane with Urquhart, is the photograph of the rocker plane as it originally was in the days when Preston and myself tried it out by falling weights in the year 1903 at Errol.

Mr. Milne, in his undated letter to me, makes a statement that some difficulty arose at Errol and that Mr J. Heriot Bell of Belmont, John Milne's Uncle, should be asked if Preston could remove his machine from Errol to this private estate, where there was a thirty acre field. The reason for the removal was that there was too much publicity at Errol. The patent for the improvement of Flying Machines had not yet been granted to Preston.

I have made two calls on James Reid, Stationmaster, Murthly – a small station on the Highland Line. James Reid had been sixteen years Stationmaster at Murthly. He was brought up in Errol. It is his recollection of Preston coming out to Errol alone, always alone. James Reid, aged eight, along with eight or ten other young boys were always willing to push the plane out of Leys Farm Buildings to the flat space at Errol. His statement is that the boys held on to the plane while the engine was started, and he remembers the blast of the propeller, and they then watched the plane bounce along making hops the height of the pailings, but it would not rise right up. Mr Reid recollected the year was 1908, but that is not so clear a recollection as the picture in his memory of the plane hoping (sic) but not being able to rise into the sky. I believe the engine that was able to cause his No1 plane to hop was the four cylinder horizontally opposed engine with the propeller revolving between the cylinders. Enclosed is James Reid's photograph – he is an unassuming man and is very cautious of his statements.

In reply to your third paragraph on page (1) of your letter of 11/7/57 – Yes Sir, I have studied Preston's own words in *Flight Magazine* – and there is great detail in his writings. You will find enclosed published algebra studies showing the care Preston took to find out the best shape of wing to give the best lift – the shape is clearly shown in the photograph of his No2 Plane with the 3 cylinder Humber engine – and these wings are not very different in shape from some of our slower types of bombers of today. If you examine your pictures of the Wright early planes and compare them with Preston's shape of wing, you will, I think, give the prize to Preston.

You will please excuse me writing my views of these distant days. I don't think you and I agree but I might as well make my statement.

My honest belief is that if Preston had had the Wright Brothers home-made engine fitted to his first plane, he would have had successful flights instead of long hops the height of pailings. Preston's 1903 rocker plane in my opinion, was a better aircraft without an engine in 1903 than the Wrights front control and warping wing idea. The construction of an engine by those two young Wright Brothers shows their determination to leave the ground on a heavier than air machine. Preston had no skill or even an ambition to build an engine – he had to search for motor power.

In your letter of 28 April, 1957, you are particularly hard on David Urquhart of Kirriemuir, writing that Mr Urquhart's dates are hopelessly wrong, that you have had to prove him wrong on one thing after another.

Whether or not the Humber engined No2 plane flew in 1908 or 1910 is not of the utmost importance. I feel it is my duty to defend David Urquhart. Urquhart was a very great friend of Preston's, doing a great amount of work for him, particularly on the Humber engined machine. Urquhart was a very knowledgeable young man and could discuss flying technique with Preston in a way that none other could. I believe that if it had not been for Urquhart, the No1 Plane, in which he was foremost in reconditioning, would not have been gifted to the Dundee Aero Club. I therefore enclose copies of his two letters to me. There are many statements in these two letters that I am in agreement with.

I do hope in finishing this correspondence I would like to say that although you and I have proved each other wrong in numerous statements, that we can still be friends and can be, if necessary, of assistance to each other.

Last week, while flying in a Chipmunk, a full aerobatic plane, with finger light control, at 2,500 feet over Dunkeld on one of those beautiful summer evenings, the thought came to me that Preston had something to do with this pastime.

With kindest regards,

Yours sincerely,

(sgnd) Jas. Y. Watson

At first reading, this letter seems to wander away from the point to some degree and is quite repetitious. On further reading, one might imagine that J.Y. was deliberately working to make points. He remains polite, at times almost deferential, but fastens on to points such as the elevator arrangements and pushes home the point that Gibbs-Smith is mistaken – which he clearly is on this point. He uses underlining to introduce a flavour of sarcasm, e.g. " ...if you examine **carefully** the photographs..."; and his seemingly conciliatory sentence about putting himself in Gibbs-Smith's position turns into something of a jibe that Gibbs-Smith is unwilling to accept a major reappraisal of the plan of his next book in the face of fresh evidence.

The digression about Flaff Pete might have been a way of saying "you are the foremost historian, so you must know about this" and thinking "I'll tell you a bit of aviation history that you don't know": and the repetition could be J.Y. assuming a slightly magisterial tone – the teacher labouring the point to a pupil who is having difficulty with the subject! Gibbs-Smith does the same!

Such suggestions are merely speculation but it must be said that, although prepared to concede some points, he was obviously still of the opinion in 1957 that the main early events were as he described and he was not prepared to be brow-beaten by the academics.

Why then did he make alterations in the captions that he originally used in the Album that he created? He mentions in his reply to Gibbs-Smith of 15 August 1957 that in response to the points raised, he checked the correctness of the statements he made in the album of collected material. This could explain some of the corrections. One constant pen and ink deletion is the recurring statement regarding the purchase of an engine from Santos-Dumont, although he did not delete reference to a French engine. Where the Santos-Dumont reference arose in the first place is not clear. It appears in the undated letter of J. Bell Milne and J.Y. refers to 'an old balloon engine'.

The other alterations are of dates and some of these alterations seem themselves to have been re-altered and therefore have their validity significantly reduced. One could imagine that as more eye-witness accounts and more professional analysis came in, he was forced to change some of the material where his own recollection had been less clear in the first place.

There is another mystery about the alterations. The McManus Gallery have recorded the date of acquisition of the Album as 1954 and think it unlikely that he would have altered any entries after that. But the correspondence with Gibbs-Smith suggests that the Album had recently been in his possession and it seems at that point that J.Y. was still of the opinion that most of the original dates were correct.

Also the second summing up is dated 1957, so it looks as though he may have re-possessed it on a temporary basis.

Gibbs-Smith was obviously disinclined to accept any of J.Y.'s reasoning. He was however, as good as his word and dealt with 'other claimants' to the first flight in his book *The Aeroplane: an Historical Survey* (HMSO 1960) [59]. In this volume he covers the main points of the substance of the claim and sets out his reasoning for rebuttal (those used in his letter of 11 July 1957). He does this at some length and gives J.Y. due credit for effort but emphasizes his ignorance of the Wright brothers' achievements, rather diminishing J.Y.'s sapiential authority in such matters. His well-known book *Aviation* published in 1970 only mentions Watson in a footnote [60] and then only to include him in a group of (unsuccessful) claimants to be first to fly who would be dealt with in Vol II. There does not seem to have been a volume II, but in a following book *The Rebirth of European Aviation 1902–1908* (1974) [3] there is no mention of Watson at all.

In the 1960 book, however, he is clearly worried by the amount of publicity already generated – mainly in the newspapers – about the claim and appears to be seeking a firm closure on the question. He sets out the claim thus:

a) Rocker No1 was first built as a glider and made flights in 1903 at Errol in Perthshire with the aid of an assisted take-off device precisely similar to the one invented a year later at the Huffman prairie, near Dayton; and that there were two witnesses to such flights, who saw them from a distance.

b) That Rocker No1 was fitted in 1906 with a Dutheil-Chalmers motor which Preston purchased from Santos-Dumont in Paris and brought home with him. This machine is the subject of a photograph.

c) That the No1 was later abandoned, then – stripped of its engine – reconstructed as a glider for the benefit of Dundee Aero Club in 1908. This machine is shown in a photograph, and formed the curiously captioned illustration in the *Manchester Guardian*.

d) That rocker No2 was built in 1908-9, with a 3 cylinder 30-hp. Humber engine. This is also shown in photographs.

e) That Rocker No3 was built in 1913, with a 45-hp. Anzani motor, and flew in the French 'Concours de la Securite en Aeroplane' at Buc in 1914, where it won a prize. This is also shown in photographs.

He deals with these claims thoroughly and logically in terms suggesting he feels that the matter is finally settled, but it is worth noting that by 1960 they are already altered from the substance of the claims as set out by J.Y. in his letter of 18 August 1957. Gibbs-Smith's arguments, and the counter-arguments, could be set out as follows:

1   I think that witnesses did see a glider; that it was in some way given an assisted take-off; and that it made tentative "flights": but that it was a primitive Wright-type aircraft with forward elevator, built after Preston had heard of Octave Chanute's lectures and articles (in 1903) on his own and the Wright's gliders.

This seems to be based on an interview which presumably took place with John Bell-Milne in which Bell-Milne described a biplane with equal wing lengths and a front elevator. (Watson was known to the Wright Brothers, suggesting he had been in correspondence with them, but at what date and on what subject is not known). In a letter from Bell-Milne to J.Y. (undated, but about the time of – or slightly preceding – the presumed interview with Gibbs-Smith), he states that he saw a glider under construction in 1903 and describes it as having no wheels and no motor. There is no mention either way of equal wing lengths or of a front elevator. He also mentioned that at the time Preston Watson was receiving tuition from Professor Keunan. Documentary evidence from University College Dundee (UCD) archives indicates that this took place in the 1900–01 session, calling into question when Bell-Milne actually saw the plane. It is also possible that Watson continued with private tuition thereafter without matriculating. There is documentary evidence that Bell-Milne attended UCD from 1903–05 and it is possible that Watson had heard of the Chanute lectures and/or their subsequent publication (1902 and 1903 respectively).

In the correspondence quoted earlier, J.Y. corrects Gibbs-Smith on the matter of the front elevator and J.Y. was considerably involved with the machine – John Bell-Milne was a casual observer on one occasion when the construction was not complete

Putting together the witness of Alex Robertson (b. 1890) – that there was a plane at Errol in 1903/4 which was kept in a shed at Leys Farm – and that of Alec Patterson (b1892) who helped to move the plane and describes the Watson brothers adjusting the engine, describes a front propeller, and saw the machine making powered hops in July of that year; then there is some cross corroboration that there was a tractor-engine plane being used experimentally at that time at Errol.

If this is then placed alongside the statements from other witnesses who can quote date-fixing events, e.g. Kerr B. Sturrock (relates events to his marriage in 1905); Henry Band (relates the timing of events to his return from the Boer War, his marriage and his departure from Errol); John Simon and Frank Brown (who time the events from their places of employment), it is difficult to sustain the argument in favour of a Wright type glider and the later dating of plane No1. J.Y., who was involved with the first two planes, flatly denies the existence of a Wright type glider.

Apart from the Gibbs-Smith hypothesis, there is no mention what-soever anywhere else of a Wright-type design and no photographs thereof.

2. "I say 'primitive' because Preston was then a young businessman of twenty-two, without technical knowledge or experience of any kind, and could not have achieved more than a primitive aircraft".

Although phrased politely enough, this seems a little condescending.

Gibbs-Smith, in his letter to J.Y. of April 1957, mentions the point that Thomas Watson, giving evidence at the inquest, said that his son had been interested in aviation for the past seven years. This seems solely based on one local newspaper report of a question asked by the foreman. It is not mentioned in the deposition given on oath by Thomas Watson and not mentioned in the report of the *Sussex Express* published on the day following the inquest (see Chapter VII).

Watson had been sufficiently interested before he was twenty to observe and actually experiment with birds and models. In church-pew 'graphiti' in Kinclaven Church where both families worshipped, there are carvings from the Philip and Watson families dated 1899 and 1900. One of these reads "P.W. – AVIATOR [24].

At age nineteen, he matriculated at UCD for study, in physics only, for session 1900–01 (rather than a general science degree) and if, as J. Bell-Milne implies, he was still receiving tuition from Keunan in 1903 (of which corroboration is lacking), it is further evidence of his focus on achieving his aeronautical ambitions. Moreover, J.Y. apparently sent Gibbs-Smith some papers containing Watson's calculations about wing aerofoil shape during the design phase of plane No2, although these, among much else, have not survived. (The plans used for the 1909 patent application are those of plane No1)

To imply that he had no technical knowledge or experience of any kind does not seem well founded. He appears to have been very much 'hands on' in the construction of his planes; but he also had the resource of employees from his father's and his father-in-law's firms.

Reference was made in Chapter 2 to the progress that had been made in the science of Fluid Mechanics, which basically determines why an aeroplane is able to fly. Bernoulli established his equation in 1738 which enabled a calculation to determine the relationship between fluid pressure, fluid speed and temperature.

One would also understand that the camber of a wing affected the cross section over which air was passing. The camber effectively reduces the cross section of air and increases the distance the air must travel over the camber. This causes the air to accelerate. Bernoulli's equation determined that as the velocity of a fluid increased the pressure decreased. So the pressure on top of the wing was less than the surrounding air, which gave the wing its lift.

All this would be known to Watson, who had studied physics at school as far as he could take it and then by continuing his studies under Professor Keunan at Dundee. This information would thus be available to him to determine the size of wing he would need and also what section of aerofoil he would need in order to enable his plane to fly. He could certainly have had access to the published notes of Sir George Cayley and the various reviews of his work published throughout the 19th Century.

The Wrights – mainly Wilbur – accessed the mathematical and physics theories already propounded by such 18th century authorities as Bernoulli and Smeaton, possibly via the interpretation already recorded by Lilienthal. They certainly had compiled a library of such works. They had acquired knowledge from the Smithsonian, starting from a request to Samuel Langley in 1899, which led them to carry out research using information such as that of Lilienthal and others. However they later discarded this as they suspected it could be inaccurate, which led them to carrying out their own research using their well know experiments of different aerofoil sections on their bicycle research 'vehicle' and their subsequent wind tunnel research. Their experimentation was empirical and they certainly carried out a considerable number of well-recorded practical experiments in determining the section of their aerofoil using their wind tunnel – remembering that their aerofoil had only the top surface to it. This was necessary in order for the wing to warp.

In Gibbs-Smith's letter to J.Y. Watson, he makes great play on the article that Watson wrote for *Flight* magazine in 1914. However the two phrases that he quotes from the article are taken out of context. Whether he did this intentionally or not, it is difficult to say.

Watson was asked by *Flight* magazine if they might publish the account of his aircraft which he had prepared for the competition at Buc in France in 1914 where, according to Manson, it attracted much admiring attention from other competing aviators.

In starting the article Watson explains why the aviators in Europe had followed the Wrights in using the wing warping idea for their planes. The Europeans were not too concerned early on about lateral stability for their designs, and had not perceived the need to bank into a turn – but the Wrights and Watson had. The Europeans later followed the Wrights' idea of warping, because as far as they were aware they thought the Wrights were "the first to fly in a practical way" (i.e. including banking). The Wrights themselves felt that they did not achieve practical flight until 1905. Gibbs-Smith claimed that Watson had derived the rocking wing idea after reading about the Wrights' wing warping system and claims that the latter was unknown in Europe until the Wrights demonstrated their Flyer in France in 1908. This ignores the fact the Robert Esnault-Pelterie attempted to copy the wing-warping system in 1904 [**35, 61**].

Watson then goes on to explain the problem that the Wrights found with wing warping. The first Wright gliders had no vertical rudder. The wing warping was intended to keep the wings level. However, Watson states that the Wrights found that wing warping caused problems for them, because twisting one wing down and the other one up caused what we today would now call Aileron Drag resulting in an out of balance force in the yawing plane. So the Wrights had to add rudders to their Flyer to keep control of it in all three axes. The rudders were first of all fixed, then when they found that was not sufficient, they had to provide a means of turning the rudders in coordination with the wing warping. They lay prone on the lower wing to reduce drag, had a stick in front to control the elevators, pedals to control the rudders and to control the wing warping. In 1903 they devised a mechanism to control the wing-warping using a hip harness and used it for the first time in a powered machine at Huffman Prairie (near Dayton) in 1904 [62].

It is interesting to compare the structural design of the wings of Watson's aircraft with those of the Wrights. In structural terms a wing is in fact a beam that carries the load of the fuselage, engines, passengers, etc. The support for the wings is the lift forces that are generated by the wings passing through the air.

The Wright Flyer was a biplane with the upper and lower wings acting as the upper and lower flanges of a deep beam, held apart by solid struts and the diagonal wire bracing. The Wrights were concerned about banking the aircraft into a turn. To achieve this, their solution was to use wing warping; one wing would be twisted downward and the other upwards in an effort to lean the aircraft into a turn.

However, in order to allow this, the wings had to be relatively weak in torsion so the surfaces of upper and lower wings were made from fabric. Having an upper and an under surface covering would turn the wing into a much stiffer torsion box, and make it impossible to twist. The Wrights used a harness attached to their hips to twist the wings into a turn.

In order to be a structural beam, each wing relied entirely on the strength of the main wing beam running span-wise through each wing and there was very little contribution to structural strength from the fabric on the upper surface.

Watson's wing design was quite different. For a start his aircraft was not a biplane. The upper wing was not attached to the lower one. Watson's solution to banking and turning the aircraft was quite different. He tilted the upper wing, much the way that a hang-glider does today, to turn the aircraft. There was no need to deform the wing which behaves as a cantilever wing – which was free – supported and held at the mid-span point on pinned supports which enabled it to pivot up and down on either side.

Watson's wing was probably covered on both upper and under surfaces, more convincingly seen on photographs of plane Nos 2 and 3. This would enable the wing to have a stiff torsion and, compounded with a relatively deep main wing beam, the upper and lower surfaces would give a strong structural strength to the wing. The lower wing in plane No1 was a cantilever wing also with no support, but plane No3 clearly has a diagonal strut propping the wing out to approximately two-thirds of the span.

So Watson was again considerably ahead in aircraft design by designing a cantilever wing, a design solution which did not come into general practice until the mid thirties when the likes of the Hawker Hurricane and Supermarine Spitfire were created.

The more one looks at Watson's design solutions, the more one is amazed at the uniqueness of his aircraft, which seems to have taken little or nothing from other designs of the time.

So Watson next explains in his article how simple his system was using his 'rocker' wing, the upper wing. He had obtained this from watching gulls soaring in his youth. This rocking upper wing tilts from side to side in order to turn, the same way as a modern Hang Glider does and the great beauty of Watson's wing was that he had **no** Aileron Drag. As the wing was tilted, the lift component remained vertical to the plane of the wing. This resulted in a vertical component still for the aircraft and a lateral component which pulled the aircraft round. This was controlled by a stick hanging from the 'rocker' wing in front of the pilot and at the same time it also controlled the elevators at the rear of the aircraft. So Watson had one control in front of him which controlled the aircraft about all three axes. It really was a most ingenious device.

Gibbs–Smith makes no recognition of the ability that Watson had as far as aerodynamic design was concerned. J.Y. refers to the 'algebraical graphs' that Watson had produced in determining his wing section. The Wrights on the other hand decided to reduce the camber in their Flyer in order to give it better lift, which is quite contrary to what any scientific calculation would give, suggesting they had not understood the scientific principles arising from Bernoulli's equation for fluid mechanics. Watson, on the other hand, increased the camber in plane No2 in order to increase the lift and thereby improve the take-off capability.

The Wrights did a brilliant amount of work in determining their design, but Watson was no less a brilliant designer himself. His 'rocker wing' came from nowhere other than himself and there is no other reason to doubt that his whole design came from his scientific knowledge combined with his studies of nature and the way that birds flew.

There is no evidence whatsoever that Watson waited to see the Wrights Flyer over in Europe before he began his design (there is evidence that it was already in his mind, if not already constructed, in 1903); and no evidence that the Wrights' European demonstrations would influence his thoughts in devising his very clever and unique way of turning an aircraft and achieving lateral stability.

To imply that at twenty-two years of age he was too young to conceive a good or original idea is obviously not tenable. For Gibbs-Smith to use this as an argument is surprising as he need look no further than Santos-Dumont for an parallel example; to say nothing of George Weir who became a director of the Glasgow engineering firm at the age of twenty-one and had completely turned its fortunes around and become the managing director by the age of twenty-five; or Watson-Watt who was 'isolated in Scotland', but under the tutelage of Keunan's successor, graduated from University College, Dundee in physics. At the age of twenty he conceived the research which had progressed to the development of what was to become radar by the time he was twenty-five years old. It would be reasonable to add to this list the name of Kerr B. Sturrock who would have been in his early twenties when he devised the laminated propeller from first principles and practical experiment; of course, this last could easily have been a lesser known fact of aviation history, overlooked at the time Gibbs-Smith was writing.

In this context, it is worth noting that sometime around 1995, Pat Peebles had the idea of applying the paddle principal to aviation. In 1998, he published his initial design in *Model Aeroplane News* [63]. He constructed a flying model and patented it in 1999.

The idea was taken seriously by the industry and was reported as being under active development in 2012 [64]. This 'Fan Wing' is much more sophisticated than Watson's undeveloped idea but the underlying principle is similar. Pat Peebles, although not a young man at the time, had no previous aviation expertise when he hit on the idea; and by that time there was nigh on a century of fast developing expertise and experience upon which to draw.

This seems to further weaken Gibbs-Smiths arguments: firstly that someone with no previous experience or training in aviation could possibly design an aircraft of the ingenuity of the Watson machines; secondly that his 1907 patent was not so ridiculous after all (although, of course, it would have seemed so to establishment thinking in the 1950s and '60s).

3. Gibbs-Smith argues that **only** the Wright brothers used skids and therefore Watson **must** have copied this feature once details of the Wright glider were exposed. However, the use of skids was/is common practice in agriculture for overcoming the problem of transport over soft or uneven or snowy terrain.

Watson was operating over such terrain and could draw upon the experience of generations of farmers in Scotland! The Wrights might well have selected skids for the same reasons as they were operating over sand. In the final analysis, there are only three possibilities – skids, floats or wheels, and all three were in general use well before the advent of aeronautics.

4. Referring to the postulated Wright-type glider, Gibbs-Smith goes on to say: "It was, I think, this machine which Preston kept by him until

1906, realizing, as we now know, that it was impossible to purchase up to 1905 or 1906 a petrol motor of a power-weight ratio anywhere near favourable enough to allow of powered flight, to say nothing of the severe difficulties in designing and making an adequate propeller".

It is probably true that such engines could not be purchased in 1903, although they existed – the Wrights built one in 1903 and Charles Manly produced one for Langley in 1903, which the Ariel Experiment Association (AEA) subsequently showed to be adequate [65]. But it was quite possible that Watson tried with whatever motor he could obtain and there is eye-witness evidence that he was trying with powered flight in 1903. Perhaps he **did** purchase the Chapin 'double De Dion Bouton after all. Discussion about the identity of the early engine/engines is dealt with in more detail in Chapter IV.

Gibbs-Smith makes no mention at all of Kerr B. Sturrock's work in making propellers. He was at it for some time and according to his own account had evolved a practical propeller by 1905 (he related the times to the date of his marriage). This was based on antecedent work with different woods which J.Y. remembers being smashed by the 'little French engine'.

5. "I believe that in 1906 it was Preston's Wright-type glider – possibly improved meanwhile – which received the engine which Preston went to Paris to buy... There is ample evidence that he did buy an engine in Paris, but we do not know of what make: the claim that it was a Dutheil-Chalmers can be disproved, as that firm did not start building motors until 1907. The newly engined glider was certainly a failure in this year 1906, and no news of it has survived. Even Santos-Dumont, with all his ingenuity and his efforts with two successive and powerful Antoinette engines, could only just get his '14-bis' machine off the ground, late in 1906".

All this is predicated, of course, on the existence of a Wright-type glider which must have been engined by a French engine other than a Dutheil-Chalmers and that plane No1 was not even built until 1906 or later and was then fitted with a (post-1907) Dutheil-Chalmers engine (or engines). One would also have to accept that Preston Watson would make no reference at all to an experimental machine or what had been learned from it; and ignore the fact that J.Y. flatly denies its existence even though he were present at the time. (As indicated earlier, J.Y.'s version was that Plane No1 was the original one – and the numbering system supports this – which would have started with a "little French engine" and been fitted with the "curious" engine later (i.e. the double Dutheil-Chalmers). There is no reference and no illustration which supports the existence of such a machine and it is therefore entirely a surmise on the part of Gibbs-Smith in attempting to order the reported facts and items of evidence into a sustainable account.

Moreover, Gibbs-Smith's time scale compresses the life-history of plane No1 if one accepts that Urquhart left Dundee in 1910. It would have to be fitted with the engine (if manufacture started in 1907, this is likely to be c 1908), test-flown, discarded, stored for some time, donated to Dundee Aero Club and restored in the members' spare time and flown by Urquhart and others as a glider – all before 1910.

W.R. Gibbs [31] stated that plane No1 was delivered to the Aero Club premises in the Nethergate by lorry, was stored there for some time and removed piecemeal to Gourlay's Yard as the restoration proceeded.

The point about Dutheil-Chalmers starting manufacture in 1907 is clearly wrong. The history of the Dutheil Chalmers Cie. is dealt with in Chapter IV. It is very likely that the Dutheil Chalmers units were two 8hp models. As mentioned Santos-Dumont had purchased two which were fitted to his No11, indicating that the engine was available before 1905.

The argument regarding the '14-bis' is perhaps irrelevant. This machine was much larger and presumably heavier and it is impossible to know the comparative aerodynamic performances. Certainly, either Antoinette engine, being water-cooled would be much heavier. Eye witness accounts describe Preston and J.Y. lifting their plane over a fence without assistance.

6   Referring to the photograph of the reconstructed glider taken at
     Gourlay's yard with David Urquhart (not Watson as Gibbs-Smith
     originally thought) in the pilot's seat, Gibbs-Smith writes "The
     photograph shows conclusively that this is indeed the No1 Rocker,

dismantled and reconstructed; but the Dundee Aero Club was founded, not in 1908 as stated, but in 1910, and that is undoubtedly the date of this glider reconstruction".

This is a reasonable point to make and it may be that Gibbs-Smith had information on the Club from an unpublished source. The first mention of it that can now be produced is in the Dundee Directory of 1910–11 [66], when the premises were at 31, Nethergate. David Urquhart was treasurer and secretary and the entry remained the same in the 1911–12 editions. In the 1912–13 edition the treasurer and secretary was J.A. Farningham and by the following year the premises had changed to 10, Constitution Road, premises occupied by the YMCA. According to W.R. Gibb, it was the earlier premises in the Nethergate to which plane No1 was brought for reconstruction. There is a press report that the Club was founded by Preston Watson [67] and one might expect that at that time he might have been made the first president, or at least an office-bearer, but he was not listed in the list of committee members or office-bearers in any of the directories mentioned. In the 1910–11 directory – which must have been prepared in 1909 – the President was William High, the vice president was D.C. Thomson, David Urquhart was treasurer and secretary and his ex-employer David Shaw was on the committee. William High (later Sir William High, Lord Provost of Dundee from 1923–29) and D.C. Thomson, owner and managing director of the nationally famous publishing and newspaper company, were prominent citizens at the time and it is surely likely that the club would have had to grow and exhibit some degree of achievement and/or permanence before they would agree to head it.

On 10 December 1957, Urquhart wrote in a letter to J.Y. "It was the second plane with the 30hp Humber engine that Preston and yourself were making flights on, when the Dundee Aero Club was formed. That would be 1908". Urquhart was the founder and first secretary and is likely to be correct. Moreover, Urquhart is unlikely to be mistaken about the date of his marriage and career move to Edinburgh. Like some other members, he may well have retained his involvement after his move. Certainly there were members as far afield as Loch Lomond.

7. The claim that "the Humber-engined No2 rocker was built in 1908–09, was disposed of when the photograph was examined, as that engine – which is clearly identifiable – was not on sale until 1910." This is unarguable (see Chapter IV) and does call into question the reliability of some of the eye-witness statements. It also corresponds to the eye-witness account of the stationmaster from Forgandenny and the article in the *Advertiser* of October 1909.

8. Gibbs-Smith accepts as correct the date of 1913 as the date of construction of plane No3; and this accords with Manson's statement that it was built to enter the 'Concours de la Securite en Aeroplane' at Buc in France in 1914. The competition awarded prizes for the safety devices and although some of the newspaper articles about Watson's achievements stated that he won a prize for the rocking wing device, it seems that this was not so. Manson, quoted in an interview with *The Peoples' Journal,* dated 7 May 1960 [68] stated that Watson "was awarded third place in a competition for an invention to counteract sideslip" and considered that the winner of the first place was not worthy of a prize as the pilot concerned admitted that he had never tried it out and "cut the thing out" once he had the plane in the air as he "felt safer that way". (It was some sort of ejection seat device).

Manson had travelled to France with plane No3 in March 1914 and stayed with it for the four months it was there. He flew the plane on a number of occasions (but omitted to tell his wife!) and on one occasion crashed it, but with no damage to himself and repairable damage to the plane. He claimed that it was much admired by some of the other pilots there. It is, of course, possible that he may have been misquoted on the question of a prize.

It is perhaps now worth examining, point by point, Gibbs-Smith's reasoning and evidence used in his rebuttal. As can be seen a substantial proportion was conjectural without hard evidence. Coming from an experienced and well recognised academic aviation historian, this carries a good deal of weight. However, it is worth examining the other possibilities without denying the items of 'hard' evidence.

It seems at least *possible* that Watson was flying in a similar way to the Wright brothers in 1903, but it was *their* flying performance at that time which became the yardstick of what constituted a flight, but there is no evidence that could support a prior claim to controlled and *sustained* flight by Watson before that. That, of course, does not obviate the possibility that he was making what would now be defined as 'hops' about that time, as eye witness accounts report.

These arguments by Gibbs-Smith are presumably what induced J.Y. to alter some of the dates ascribed to his version of events and to the dating of some of the photographs. But he apparently was never convinced of the Gibbs-Smith version of events and it has to be remembered that J.Y. was present at many of the events. But the rebuttal has been accepted historically.

Having quashed the claim, Gibbs-Smith, in the same section on Preston A. Watson in his 1960 publication *The Aeroplane; an Historical Survey*, goes on to deal with the invention of the rocking upper wing device. The language he uses seems to be detracting from the invention and belittling of the achievement. This may be unintentional but is in line with a flavour of sycophancy in relation to the Wrights that appears elsewhere in Gibbs-Smith's writing. It is worthy of note that Gibbs-Smith, in the context of this argument, does not bring out the fact that the Wrights' Flyer could not fly without an assisted take-off. In December 1903 this was a downhill start down the launching rail into an existing windspeed of 27mph. Their falling weights and derrick device was first used at Huffman Prairie in 1904 and it seems still to have been in use when the Wrights came to Europe in 1908. There is a view (mainly amongst the French) that the accolade should go to Santos-Dumont who did make an unassisted powered take-off in November 1906!

Gibbs-Smith writes "Preston, in his 'Flight' article of 15 May 1914, makes no mention at all of his No1 Rocker, nor of any efforts to build or fly it at that time, let alone as early as 1903; he only included photographs identifying 'Watson No2' and 'Watson No3': he clearly must have thought of the No1 only as a tentative prototype, which indeed it was. He then demonstrates quite clearly in the article that his rocking wing idea was specifically conceived as an alternative system to the Wrights': "the method of preserving lateral equilibrium invented by the Wright Brothers has been slavishly followed", he said, "but this has probably been due to the fact that these gentlemen were the first to fly in a practical way. It remains to examine whether it is not possible to invent a method of preserving lateral equilibrium, which requires small power on the part of the pilot, etc., etc.: he then proceeds to describe his idea of a rocking wing aircraft, and tell of its operation and his experiments. It would have been quite impossible for him to understand the proper workings and implications of the Wrights' control system until after August of 1908: this was the date when the public first saw a Wright aeroplane, when Wilbur Wright started flying at Hunaudieres in France, and thereby revolutionised European aviation. It was then, also, that the Wrights' launching apparatus was revealed and could be studied for the first time: a weight on the end of a rope…"

"Having studied the Wrights' aircraft and its system of control, and the method of launching – which was soon widely described and illustrated – I believe that Preston then thought of his rocking wing idea, which he felt was a practical alternative to the Wrights' system. He then built his rocker No1 on skids – like the Wright machine – and tried launching it by the method just described, except that he saved expense by using the bough of a tree in lieu of a derrick."

Once again, Gibbs-Smith's phraseology conveys an absoluteness which engenders a suspicion of a biased view. Terms used in the numbered paragraphs 8, 9, 10 of his letter to J.Y. of 11 July 1957 are illustrative – "it is surely quite obvious.." "there cannot be any doubt about the origin and history of Preston's rockers", "the rocker idea cannot have anything to do with whatever happened in 1903–4" and of the eye-witnesses – "You will agree that they are very vague".

These may not be very strong or even valid points on which to argue matters and it risks investing Gibbs-Smith with an attitude he may or may not have adopted, but they do influence the way in which one deals with those components of the discussion which are based on opinion, rather than evidence – i.e. most of it!

He claims 'virtual proof' by citing the date of the 1909 patent and the fact that the plane used to illustrate that patent was the No1 rocker.

But Gibbs-Smith is unlikely to be correct on the issue of Watson conceiving the rocker idea after 1908. If one reads the detail of his 1907 patent, his treatment of the idea for lateral control by means of a rocking wing indicates that he had already conceived it, and very probably implemented it. The complete specification of the 1907 patent p4 line 50 et seq. (Chapter V, p. 50) mentions the rocking wing device. The material for the application must have been prepared in 1906/7 and he is likely to have had the first-hand experience from plane No1 beforehand.

Oliver Stewart, in his article of 1954 [58], wrote "I have heard only one objection to the Watson claim and that is: Why was it not heard of earlier? That seems a curiously weak objection when it is a matter of history that only three American papers thought the Wrights' first flight was worth reporting, and that nobody knew anything about it or had heard of it until several years afterwards.

Pioneers do not always realize that they are pioneers and that is why some early achievements are short of documentation and witnesses. I personally was well impressed by the Watson claim, though I would not attempt to put a date on the flights. But I feel that Mr Watson did well to attempt to draw attention to the work that was done by his brother.

Two interesting features of the Watson aircraft (whether 1903 or later does not much matter, for it was certainly pre-1909) are the tail, which is a box structure and might have been derived from the Hargrave box-kite work, and the lateral control system…"

There is also the problem with the Gibbs-Smith argument of having to discount the eye-witnesses as all being wrong:

a) Johnny Gourlay's father fetched the weights from the blacksmith and used to help with the launching tackle at Forgandenny. He also recounted the details of substituting graphite for the lard used as track lubricant. He dates these activities as prior to 1906.

b) John Harris helped with handling the plane and hoisting the weights at Forgandenny in 1908

c) John Logie, in1908 helped to handle the plane to the launching site and witnessed the launching mechanism being rigged at Errol.

d) Alex Paterson [32] describes his involvement with a front-engined machine in 1903. To be fair to Gibbs-Smith, this last was not published until 1966 after the publication of the rebuttal and was not the subject of correspondence directly with J.Y. But Gibbs-Smith must have had an excellent 'intelligence' system because he responded immediately via the columns of the *Dundee Courier* [69]. He was constrained in the length of his response in that he did so via the correspondence columns, but the terms in which it was couched were quite dogmatic. Points which were more discursive when he was dealing with J.Y. had become absolute truths, e.g. "There is no truth in the story of the 1903 Watson flight"; "Apart from all these matters, which conclusively proved the whole Preston Watson story had been unconsciously pre-dated...".

He alludes to J.Y.'s admission that he (J.Y.) made no such claim and quotes again Preston Watson's sentence about the Wrights being the first to fly in a practical way in his 1914 article, but goes on to say that he could not possibly have the skill or understanding to make a flyable aircraft. It sounded very much like a rather grumpy re-iteration of his already published reasoning – grumpy that someone would have the temerity to challenge what he had published as the final word on the matter.

There remain doubts about most of the points made in support **and** in rebuttal of the claims and it is left to the reader to construct a balance of probabilities.

James and Preston Watson are given a brief mention by Desoutter in 1954 as having built and successfully flown a glider of original design in 1903 [70]. However, Gibbs-Smith published the most widely accepted 'last word' on the matter in the 1960 volume *The Aeroplane: an Historical Survey of its Origins and Development*. In his subsequent 1970 volume *Aviation* Watson is mentioned only as a name in a list of other claimants, and in *The re-birth of European Aviation* published in 1974 there is no mention at all. Subsequent authors have either omitted Watson from the pioneer events, or re-iterated Gibbs-Smith. For example, the comprehensive book *Aviation* by Peter Almond [71] makes no mention of Watson, attributing the first European powered flight to Santos Dumont in 1906.

It may be worth quoting a paragraph from Gibbs-Smith book *Aviation*

[72]: "A passing tribute-perhaps coupled with a retrospective sigh – should be paid to the many inventors on both sides of the Atlantic who designed a multitude of aircraft during 1909. Few of them ever left the ground, few possessed true originality, and fewer still combined originality with any degree of practicality. In short, there was in 1909 – along with the few successful aeroplanes – a prodigious amount of wasted aeronautical energy."

Nevertheless, there has been a smouldering thread of articles about Watson, most of which have been quoted or referred to here; and many of which finish by regretting that he has not been widely enough recognized for what he achieved in his short life [11, 72, 73].

As well as the report of the interview with Manson in the *People's Journal* in 1960, there was an article published by the Glasgow branch of the Royal Aeronautical Society in 1966 which re-iterated Gibbs-Smith's views. In 1981, an interesting and well-produced book by Adamson and Lamont-Brown about Old Dundee [74] had photographs of Watson and Plane No2 with a covering text which contained many of the inaccuracies that had survived in the columns of the press.

The matter surfaced again in 'The Craigie Column' of the *Courier & Advertiser* in 1992 [75] and it was here that one contributor stated that Watson had founded the Dundee Aero Club. The correspondence ('The Craigie Column' consisted of contributions from readers) ran between 9 January and 5 March 1992 and contained elements of the Watson story, but also many inaccuracies. As well as a contribution from the daughter of James Manson, the correspondence in this column also contained a contribution from the son of William Dow who was a member of the YMCA in Dundee and knew Watson well. He had recounted to his son that on one occasion after he had left Dundee to work for Argyll Motors of Alexandria (near Loch Lomond), he had been asked to deliver a new car to a dealer in Dundee. "Instead of going straight to the showroom in Reform Street, Mr Dow drove the car to the YMCA to impress his friends – only to find Preston Watson going one better, showing off an aeroplane. Shortly before a take-over closed Mr Dow's department, he was able to furnish Watson with a sleeve-valve engine, to improve his aircraft. Whether or not the Argyll Motors engine was ever fitted, however, my informant doesn't know".

There is no dating on the information given and it seems likely that Mr Dow was delivering the car during the more prosperous days in the fortunes of Argyll Motors, i.e. up to 1907–8, after which Argyll Motors of Alexandria underwent liquidation. However, it re-appeared as Argyll Ltd. in 1910 and it was not until 1911 that the single sleeve valve engine of 15–20hp was developed. This seemed to be a very good car engine and was continued until 1913–14 when the company again went into liquidation. It was a 6-cyl, water-cooled engine, making it very unlikely that it could be effectively fitted to a Watson plane. It seems most probable that the exhibition at the YMCA was around 1913–14 and, although it is possible that this was his new plane No3, it seems more likely that this was plane No2, or the restored and engineless No1.

The *Scotland on Sunday* published an article by Peterkin [76], which was generally supportive of the early dating of the Watson flights which produced a tartly-worded response asserting the contrary. This contained the information that J.Y. had 'gone public' about his efforts at an RAF mobile exhibition in the City Square, Dundee in 1949 [77]. Certainly, J.D. Leslie in his column in the *Daily Record* was using J.Y.'s material in 1949 [78, 79].

Up to this point, Gibbs-Smith concentrated on the events of 1903, which involved a re-writing of the timetable of events as set out by J.Y. and centred to a major degree on the chronology of plane No1. Having put forward the above case for his beliefs, he had to offer an adjustment of the dates and events regarding planes No2 and 3. As has already been discussed, a number of those witnesses involved at the time stated that plane No2 was flying in 1908, but there is the hard evidence that the plane had a 3-cyl. Humber engine which could not have been purchased before late 1909 at the very earliest. There is the remote possibility that there was another engine fitted to plane No2 before the Humber engine, although no records anywhere make any mention of this. If it were that he initially used the engine from plane No1, this could explain why the patent drawings showed the Dutheil Chalmers engine. Even this seems unlikely as the undercarriage illustrated was that of plane No1.

According to Manson, plane No3 was under construction already in 1913 and the aim at that time was to have it ready for the *Concours de Securite en Aeroplane* at Buc, near Paris. Gibbs-Smith does not dispute the fact that it was there but does correctly refute J.Y.'s claim (re-iterated in subsequent newspaper articles) that the rocking wing invention earned a prize at the Concours. The statement may have started from an enthusiastic interpretation of Manson's statement that he took third place in the contest. There may only have been one prize in each category.

Gibbs-Smith states, "it won neither of the two main prizes awarded, nor any of the consolation prizes. The magazine *Aeroplane* (17 July 1915) commenting on his final No3 Rocker, said 'the machine was not a success, partly no doubt, as Mr Watson claimed, because of its being underpowered. Nevertheless it did get off the ground for short distances'."

The above quote from *Aeroplane* [80] is accurately quoted, but perusal of the whole short article reveals that it must have been written from an office desk and a reporter's notes – certainly not by anyone who was present at Buc. It is primarily a report of Watson's death, amongst columns devoted to casualties and fatalities in the various theatres of war. The sentence quoted is part of a short 360-word well-intentioned overview of his life. There are some inaccuracies and it sounds as though the sentence in question might be based on remarks about the earlier machines. Certainly plane No3 was not underpowered.

The suggestion that it was not a success and hardly managed to fly, let alone compete for safety devices, is at diametric variance with what Manson (who was there) reported later. He described how many of the other pilots there thought well of it.

Manson's account is not easily dismissed. He was deeply involved in the building and maintenance of plane No3. He had flown it himself; he was in charge of it during the four months it spent in France; and, as already mentioned, when war broke out he enlisted, initially in the Black Watch, but soon transferred to the RFC and finished in charge of engine fitters at Peterborough. His experience and insights were well founded.

Flight magazine (3 July 1914) [81] has pictures of the competing machines at Buc. The caption for the only British entry reads "The only British machine entered for the competition – The Watson rocking wing aeroplane. We are told by Mr Somerfield of Melton Mowbray, who piloted this machine, that the "Watson" was ruled out of the competition for no apparent reason, as it flew quite well once Mr Somerfield got used to the rather novel control".

Based on the French press coverage of the event in *L'Aerophile* [82] and *La Revue Aerienne* [83], it seems that the competition was a protracted affair running from 1 January to 30 June 1914, and involved two airfields – Chartres and Chalons. Judges were appointed by *l'Union pour la Securite en Aeroplane*. The chairman, M. Lecornu, was a member of the Institute; one of the two vice-presidents, M. Soreau, was the vice-president of *L'Aero Club de France* and there was a Secretary, M. Savary. The remainder of the panel of Judges included three from the War Ministry; one from the Ministry of Marine; one from the Ministry of Public Works and a further seven from the Union de Securite en Aeroplane.

The panel numbered sixteen. Their brief was to examine the workings of the various devices entered under conditions resembling the conditions of flight and, where appropriate, to witness the actual performance in flight, executing specified manoeuvres. Parachutes were examined apart according to a separate protocol; aircraft fitted with stabilizing devices were demonstrated in flight; ancillary apparatus to engines (e.g. carburettors) were demonstrated attached to aero engines.

On 25 May the Jury divided into two groups – group 1 assembled at Chartres and group 2 at Chalons. It seems there were originally 49 contestants, but only thirty-one drew lots for allocation to groups 1 and 2. Contestants had to arrive at the allocated airfield by air between 07.00 and 09.00 and were expected then to carry out a series of tests:

i)   Taxi and take-off in as short a distance as possible, circle to the left and make a glide approach and landing from 200 metres.

ii) Execute a circle to the right, make a landing with the engine running and stop in as short a distance as possible.

iii) Fly from Chartres to Chalons or vice-versa, stopping en route to refuel at the pilot's discretion.

Watson was allocated to Chartres.

On the morning of the 25 May, only four aircraft turned up. These were a Bleriot monoplane, a modified Bleriot monoplane, the Caudron Brothers' biplane and the Schmitt biplane. During the first two tests the Bleriot monoplane crashed on the glide approach. The third test, the cross-country flight, was completed only by the Caudron. The pilot of the modified Bleriot got as far as Buc but did not continue. The Schmitt landed 30km short because of engine trouble.

Wind speeds were 10km/sec gusting to about 15km/sec. This presumably hampered the operation and may have explained the poor turnout; although *l'Aerophile* suggested that many of the contestants simply did not have their inventions at the stage where they could be assessed.

The contest was re-scheduled for 11 June at Buc. The first two tests were as before but the cross-country flight was reduced to a flight from Buc to Chartres. Referring back to the caption from *Flight* magazine, some light was shed on the Watson entry by a report of the second staging of the event in a report of the judging published in the July edition of *l'Aerophile* [84]. Watson is unlikely to have been present the whole time and the plane certainly was flown at times by others, including Manson. It seems likely that only licensed pilots would be allowed to fly in the competitive events and Watson did not at this time have a licence. Apparently the usual pilot of the Watson machine was replaced before the repeat competition and although it was reported that the new pilot was able to roll the aircraft and land it, he did not perform the tasks set by the rules of the competition. Whether the original pilot was Manson or someone else from Scotland is not recorded and presumably the Mr Somerfield mentioned in the *Flight* report was a hired substitute. He regarded the disqualification as "for no apparent reason", but it may be that he was unfamiliar with the required programme, or unable to execute it.

The judging panel were interested in several of the entries but eventually decided not to award the 'grand prize' of 400,000 francs. Instead, they awarded 50,000 francs to the Sperry Company (gyroscopic stabilizing device); 30.000 francs to Paul Schmitt for an aircraft whose wings could have their angle of incidence altered by up to 12 degrees. In addition they awarded 'bursaries' ranging from 1,000–15,000 francs to a number of researchers to encourage continuing research.

As far as the Watson story is concerned, it is regrettable that the sequence of events as recorded in the history of pioneer aviation have given primacy to the dating of the early flights at the expense of the wider picture of what Preston Watson actually – and indisputably – achieved.

CHAPTER VII

"The Last Chapter"

At 05.00 hours on 30 June 1915, Watson was ready to start the cross country training flight that would complete his training requirements at Eastchurch. The route was from Eastchurch to Eastbourne.

At 05.00 the weather conditions had been such that he thought it wise to postpone departure. A review at 06.00 was no different, but at 07.00 the conditions had improved and communications with Eastbourne reported suitable conditions at that end.

The aircraft was a Caudron GIII – a type once used in battlefront reconnaissance, now relegated to a role as trainer in many establishments. These machines were cheap to purchase and had a reputation for ease of handling and reliability. Designed and built by Gaston and Rene Caudron who started the factory in France in 1911, this biplane model, powered by a variety of engines from 60 to 100hp., had become obsolescent and no match for the faster aeroplanes that were emerging on both sides of the conflict [85].

The actual plane in question was a relatively new Caudron GIII, No 3266, fitted with a Gnome 80hp engine. It was reported to be in sound condition but had been involved in an accident on 12 June when a skid had broken on landing. Watson was not the pilot in that incident but the plane had been flown on a number of occasions in the preceding week by Watson himself [25, 86].

The course he would fly would be 210 degrees (true) and the distance in the region of 60 miles.

At 07.55, Fred Jarvis, a gardener, was standing with a friend, Jesse Durrant, beside a high hedge adjoining Dunlye field opposite the Cross-in-Hand Hotel, East Sussex.

Both men heard an aircraft approaching from the direction of Mayfield. Their view in that direction was obstructed by the hedge under which they were sheltering from the rain. There was a sudden loud noise like an explosion, the engine noise ceased abruptly and debris started to fall. Almost immediately an aircraft fell through the clouds. The engine hit the ground first and the remainder seemed badly smashed. At that juncture, they were uncertain whether or not there was a man in the machine.

The two of them ran to the spot where it had hit the ground and saw that there was indeed a man in the wreckage. They ascertained that he was not moving or breathing and concluded that he was dead. Uncertain what to do, they made no attempt to move him, but observed that the damage to the plane was severe and the man was no longer securely attached to it by the straps.

PC No29 Frank Biddlecombe, the village policeman, did not have to be sent for, as he was in his garden at the time and hurried to the scene. He had heard the plane approaching and heard the same loud report, which he too construed as an explosion. He then witnessed the plane falling out of the low cloud to crash about half a mile away. Other villagers had also seen or heard all or part of the drama. These accounts more or less corresponded, but those who saw it fall from the clouds described a wobbling descent with a loud droning noise.

Dr Thomas Holman, the local doctor was called and pronounced the pilot dead. The body, that of Sub-lieutenant (temp) Preston Albert Watson, RNAS was carried to the Cross-in-Hand Hotel, where Mr J. Herring, the proprietor, made appropriate accommodation available.

At 10.30 Squadron Commander Philip Shepperd RN arrived from East-church and examined the wreckage in conjunction with Flight-Commander Douglas George Young, RN of the Royal Flying School, Eastchurch and flight-lieutenant R.H. Jones, RNAS, Eastbourne. Arrangements were made for the remains of the aircraft to be removed to Eastchurch.

The Coroner, G. Vere Benson, was informed and issued an instruction to PC Biddlecombe to assemble "four and twenty good and lawful men of discretion and understanding…" to form a jury at the inquest which would be held at the Cross-in-Hand Hotel at 2.15pm on 1 July 1915.

Apart from those already involved as witnesses or officers of law, thirteen other names are listed, viz. John Barnett Newnham; Noah Eastwood; Thomas Atkins; Harry Andrews; Harry French; Robert Jarvis; Robert Southgate; Charles White; James Herring; William Seamer; Sidney Chapman; Benjamin Hearn; Joe Hayward. James Herring, who was the publican of the Cross-in-Hand Hotel was selected to be the Foreman of the Jury [86].

At the inquest, the Coroner heard the accounts of what had been seen by the eye-witnesses.

Dr Holman's report stated that he had pronounced the pilot dead and went on to say "I think almost every bone in his body is broken and death must have been instantaneous". No post-mortem examination was proposed.

Watson's father, Thomas Watson gave a deposition on oath in which he stated: "I saw him alive in April – in Scotland. Absolutely good health. Was a keen aviator – built three machines himself. Studied theoretically – was very keen". (**Note: there is no mention of his interest only going back seven years**)

His mother did not wish to give evidence and made no deposition.

On questioning, Squadron Commander Alexander Ogilvy established the details leading up to the take-off from Eastchurch earlier that morning. At 05.00 on the 30 June, Watson was due to take off on his cross country flight for completion of his flying training. This was postponed because of the prevailing weather conditions and postponed again at 06.00 for the same reason. By 07.00 weather conditions were considered suitable. Eastbourne reported similar reasonable conditions. The plane had been checked over by Ogilvy and found satisfactory; and the flight was approved. Ogilvy considered that Watson was the best flyer among the current group.

Shepperd described his findings. The debris was scattered over a wide area, some pieces being found in adjacent fields. One piece of metal was seen by an eye-witness to fall in an adjacent wood but this was never recovered or identified. As the noise suggested an explosion, he thought one possibility could be that a cylinder burst and the shrapnel flew back to hit the wing structure causing it to collapse. There was, however, no scorching or signs of burning. The structure of the machine was badly smashed. The control wires were intact and correctly connected. The engine appeared intact but no valid comment was considered possible without fuller examination.

When asked directly about the cause of the accident, he suggested that a wing had collapsed in flight. He had seen and heard this happen on previous occasions and considered that the noise of it could be mistaken for an explosion or gun-shot.

Flight Commander Young thought it likely that Watson had become lost, had descended steeply and on glimpsing the ground closer than anticipated as the cloud thinned, had pulled back abruptly on the stick resulting in excessive strains on the wings.

No other explanations were forthcoming and with the agreement of the jury the Coroner summed up as follows:

> Killed by falling with the aeroplane he was flying – from Eastchurch to Eastbourne – owing to some accident to the aeroplane – ACCIDENTAL.

The theory that he had become lost (and perhaps disorientated) in cloud is a feasible one. He was approximately twelve miles to the West of his intended track and was probably flying parallel to it, if the observation was correct that he was coming from Mayfield (which lies approximately NNE of Cross-in-Hand). There is now no way of knowing the wind direction and strength. The precise cloud conditions likewise cannot now be known, but the cloud base was described by the lay witnesses as low and it was raining.

The death was fully reported in the Dundee press.

The Dundee Advertiser published the following obituary [87] on the morning of 1 July.

DUNDEE AIRMAN KILLED
Fall of 1,000 feet
Tragedy Supposed Result of Engine Trouble
Their Majesties' Sympathy

Flight Sub-lieutenant Preston A. Watson, the only Dundee officer of the Royal Naval Air Service, was killed by a fall from an aeroplane at Cross-in-Hand, near Heathfield, Eastbourne yesterday morning.

The accident took place about eight o'clock and although the cause is not definitely known, it is believed to have been due to engine trouble. Before the machine was seen a loud report was heard and the aeroplane fell to the ground at great speed. It dropped like a stone into a field nearly opposite the Cross-in-Hand Hotel belonging to Mr J. Herring and was smashed to atoms. Several people from houses nearby and some haymakers were quickly on the scene, but found the pilot beyond all aid. His body was buried beneath the debris of the shattered biplane, his head and shoulders only being visible from between the engine and propellor. He was quite dead. Dr Holman was called but could do nothing and the body was removed to the coach-house of the Cross-in-Hand Hotel. The wreckage was immediately taken charge of by the military authorities and the field guarded.

AN EYE-WITNESS ACCOUNT

Mrs Dove whose backyard adjoins the field where the machine fell, told a Courier representative that about 8am she heard a noise as of an aeroplane engine but on looking into the sky could see nothing. There was a loud report and then the machine was seen to be falling rapidly. After it struck the ground Mrs Dove said splinters flew into the air to a height of about fifty feet.

Other eye-witnesses stated that they heard the sound of an engine coming from the direction of Mayfield. The machine was not visible being apparently above the clouds which were rather low but it seemed to travel as far as the Cross-in-Hand Mill and then turn. Suddenly a loud bang was heard and the machine was seen to fall through the cloud. One gentleman said that after the first report the machine seemed to be planing down when there were three successive bangs and the machine fell, nose first.

After the aeroplane had landed, bits of it continued to drop, which gives the impression that an explosion had occurred. The coroner for East Sussex, Mr J. Vere Benson has been informed of the occurrence, and has fixed the inquiry for today at 2.15 in the Cross-in-Hand Hotel.

ROYAL SYMPATHY

Lt Watson, who was thirty-four years of age, was the younger son of Mr Thomas Watson, Balgowan and son-in-law of Mr Joseph Philip, Dalmore and leaves a widow and two children who reside at The Retreat, Perth Road. Mrs Watson has received the following message of sympathy from Their Majesties

BUCKINGHAM PALACE

The King and Queen deeply regret the loss you and the Navy have sustained by the death of your husband in the service of his country. Their Majesties truly sympathise with you in your sorrow.

A Skilled Aviator

The deceased officer was well-known in the city. After receiving his education at the High School he embarked on a business career in his father's firm – Messrs Watson and Philip, produce merchants, Dundee and Aberdeen of which he had been for several years a partner. From boyhood he had shown a special bent towards mechanics and during the last seven years he devoted a large part of his leisure to the study of the science of aviation and to the construction of flying craft. His ideas on the subject took original shape, and he built more than one aeroplane. while he patented what has been found to be a valuable improvement in steering gear and stability apparatus. Since the commencement of the war, his strong desire had been to be of service to his King and Country, and his inclination leaned towards that arm of the services in which he had always been deeply interested, and with which he was most familiar. In the early spring of this year, therefore, he proceeded to Hendon and in the principal training school there obtained his certificate as a skilled aviator.

In the parchment he was described as the best pupil who had ever been passed through the hands of his instructor. On 29 April he was gazetted to Sub-Lieutenancy in the naval wing of the flying corps and since that date he had been stationed in the South of England. For a number of years he was a Lieutenant in the 1st Volunteer Forfarshire Artillery.

Of splendid physique Lieutenant Watson as boy and man distinguished himself in athletics being the holder of many important trophies in different branches of sport. He was a splendid gymnast and for several years was an exceedingly well-known figure in the rugby football field proving himself a reliable threequarter back in many keenly contested games. The news of his untimely death yesterday caused a painful shock to his many friends and acquaintances in the city and the sincerest sympathy is extended to his young wife and children and his other relatives in their tragic bereavement.

One odd fact is that this was reported in the edition that came out the morning that the inquest was due to be held and therefore must have been from press interviews with those to be called as witnesses.

The matter of Watson's activities in actual flying machines going back seven years is mentioned, suggesting that Thomas Watson had made that statement to a reporter before the inquest. But the same article states that he had always been interested in aviation. The information about the accident must have been telephoned to Dundee on the day of its occurrence.

The Dundee Year Book [88] printed the following under the heading 'WAR':

30 June PRESTON WATSON, AVIATOR, KILLED.

There was an article in the *People's Journal* [89] (also a DC Thomson publication, like the *Courier & Advertiser*) headed:

City Airman Killed

How Preston A. Watson met his death.

The entry was a whole column, dealing mainly with the evidence from the inquest, but the last section read:

Sympathy from High School Directorate.

At a meeting of the Dundee High School Directors on Thursday, Sir George Ritchie, who presided, said that before proceeding to the usual business he thought it would be only fit that he should refer to the very unfortunate thing which had happened. Only a few weeks ago they had the privilege of congratulating a family upon the valour and conduct in the field of one of the High School boys. Today it was their melancholy duty to send their expression of sympathy to the family of another High School boy who had fallen doing his duty as nobly and bravely as though he had been in the field. He referred to the lamented death of Mr Preston Watson. Mr Watson was a young man of very great ability, and in his earlier years had taken to studying aviation. When this branch of the service came prominently to the front Mr Watson offered his services to the Country and was accepted, and he proceeded to London to perfect his education so that he might be fit to render these services, which no doubt could have been very valuable to his Country. Through an unfortunate accident he had come to an untimely end. The least they could do was to enter on the minutes an expression of sympathy and send it to his widow and father.

This was agreed to.

One could comment on the quality of the reporting, but once again there is only passing mention of his achievements with flying machines.

The article in *Aeroplane* [80] referred to by Gibbs-Smith was not an analysis of his aviation achievements, although these were briefly and probably inaccurately mentioned; it was a report of his death. There was a brief obituary also in *Flight* on 9 July 1915 [90] which did mention that he would be remembered as the inventor of the rocking wing system.

- Newspapers local to the scene of the accident reported the inquest. *The Sussex Express* [91] and *the Sussex Daily News* [92] both published accounts on the 2nd of July 1915. It was the latter that reported that the foreman of the jury had asked Thomas Watson about his son's previous experience with aircraft and reported him as saying that he had been a keen aviator for the past seven years.

Again, this is puzzling, but a significant point to consider in view of the fact that Gibbs-Smith uses it in support of his timetabling of Watson's experience. Why would the foreman of the jury, a publican to trade, ask this question? If the answer given to Mr Herring was the source of the press reports, how did it get to the *Courier* in Dundee before it had been given?? It would seem somewhat irregular for a juryman to be able to question a witness before the inquest; and in any case, if the accident occurred on 30 June, it is unlikely that in 1915, Thomas Watson could have arrived at Cross-in-Hand from Dundee before mid-morning on 1 July – **after** the printing of the *Dundee Advertiser* in which it appeared! It is conceivable that a Dundee reporter hurried to the bereaved household in time to interview Thomas Watson before he set off for Sussex but it seems unlikely that Thomas Watson would have granted an interview in those circumstances. However, information seemed capable of rapid travel in 1915, if the contents of the royal telegram were known before the 1 July edition went to press.

The funeral took place in Dundee with a Royal Marine Guard of Honour and he was laid to rest in the Western Cemetery there. This was the subject of a further report in the Dundee papers [93] under the heading,

FUNERAL OF LIEUT. PRESTON WATSON

IMPRESSIVE OBSEQUIES

A naval event of melancholy interest took place at Dundee on Saturday, when the remains of Flight Sub-lieutenant Preston A. Watson of the Royal Naval Air Service, who was killed in a flying accident near Eastbourne, were laid to rest in the Western Cemetery. Public sympathy in the occasion was evidenced by the large crowds which gathered at the dead officer's house, The Retreat, Perth Road, and which followed the cortege to the burial ground.

The obsequies were carried out with military honours. Six men of the Royal Marine Light Infantry provided a guard of honour for the dead flying officer, while the firing party comprised a detachment of the 5th Scottish Provisional Battalion, under Lieutenant Matley, from Dudhope Castle. Flight Commander Young, who gave evidence at the inquest at Eastbourne was present with a party of men belonging to the Naval Air service. The Brass and Pipe Bands of the 5th Provisional Battalion were also present.

The Rev. John McConnachie, St John's United Free Church, conducted a service in the house, and as the cortege began the journey to the cemetery it was headed by the firing party and the brass band, playing the Dead March in Saul. In the bright sunlight the procession made an impressive spectacle, the black clothing of the civilian mourners contrasting vividly with the khaki of the officers and soldiers and the blue dress with white caps of the naval contingent. Half way on the road the pipes struck up "The Land o' the Leal" to which air the cortege passed through the cemetery gate.

At the graveside the service was brief but affecting. Chaplain-Colonel Campbell offered a short prayer, the customary three volleys were fired while the pipes wailed out the plaintive notes of "Lochaber no More". Two buglers sounded "The Last Post" and the ceremony was over.

So ended the life and career of a remarkable young man. He left behind a young widow and two young sons.

Final Crash Site, photographed in 2008.

Mons L. Duthril

Dr M. B. Boyd.

Mons Archdeacon.

Mons Santos Dumont.

Prominent figures in French Aviation gathered round the Santos-Dumont 'Demoiselle'. The figure on the left is Queen Elizabeth of Roumania. (from 'Conquest of the Air')

*Dunne Glider, Blair Atholl, John Dunne is in the pilot's seat.
The figure with the waking stick is Col. Capper.*

Strawberry Bank.

James and Preston Watson.

Preston Watson with Beatrice.

Preston Watson on Beatty-Wright trainer.

*Muirhouses Farm, Errol where Watson began early test flights,
later forming part of the (now decommissioned) Errol Aerodrome.*

Plane No.1 at Forgandenny, circa 1906/7.

Launching tree, Forgandenny, photographed in 2014.

Model of Plane No.1 by George Jamieson.

Plane No.1 rebuilt as glider. David Uquhart is in the pilot's seat.

Santos Dumont in the basket of Airship No.1 showing the double De Dion engine.

Mechanical part of the Santos-Dumont combined Balloon and Aeroplane Machine. Two 8 H.P. D.C.B. Engines are used, weighing 25 lbs. each, complete.

*Power unit of Santo-Dumont's No.16 with twin 8hp Dutheil Chalmers engines.
These were changes later for a single 50hp antoinette,
but No.16 never flew and was abandoned in 1907. These engines might have
been available to Watson in 1906/7.*

Model of Plane No.2 by James Manson.

Plane No.2.

Planes 2 and 3 in flight. This series of four photos labelled retrospectively by J.Y. may be misleading. Top right and bottom left are definitely No.3; bottom right is definitely No.2. Top left is probably No.2, but could be either.

McManus model plane 3 by James Manson.

Bi-plan Anglais Watson by Buc, 1914.

Cross-Inn-Hand hotel.

J.Y. Watson.

Sir George Caley, aged 82 in 1855.

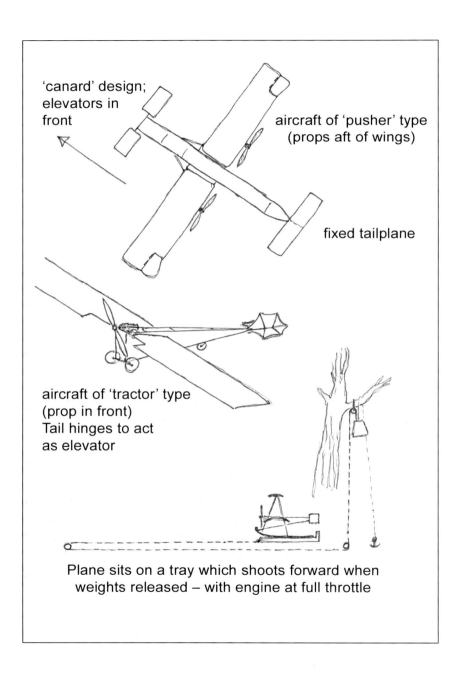

'canard' design; elevators in front

aircraft of 'pusher' type (props aft of wings)

fixed tailplane

aircraft of 'tractor' type (prop in front) Tail hinges to act as elevator

Plane sits on a tray which shoots forward when weights released – with engine at full throttle

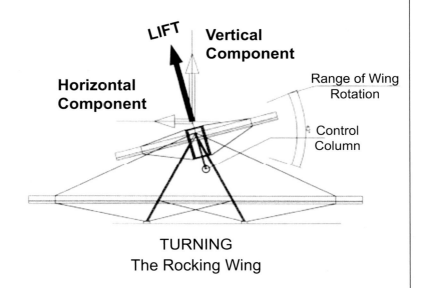

TURNING
The Rocking Wing

The principle of the Rocker Wing followed Watson's observations of sea gulls. If he wanted to make a turn to the right he tilted the Upper Rocker Wing to the right, ie right wing down.

With the wing tilted, the Lift generated by the wing would still be at right angles to the wing surface. This could then be split into a force to the right, or horizontally, as well as a force vertically. This side force then made the aircraft turn in the direction required

Preston Watson's Gravestone in Western Cemetery, Dundee.

The inscription reads: *To the dear memory of Preston Albert Watson Royal Naval Air Service who fell at Cross-in-Hand, Sussex on 30th June 1915 aged 34. Also his son Ronald Stuart Watson lieutenant RNVR killed in action at sea 26th Nov 1941 aged 33. Also Beatrice Philip loving wife of Preston Watson and mother of Ronald who died 23rd August 1971 aged 89 years. Also their son Harold Preston Watson FRCS Surg. Cdr. RN who died 9th June 1991 aged 78. Brother to Ronald.*

Postscript

Obviously this book attempts to draw together the story of Preston Watson, a fine man by any standards, who dedicated much of his young life to developing his ideas of aviation; and gave his life in the service of his country.

Progress in aviation, as in any field of knowledge, is the aggregate of the experience and experimentation of all who gain and share the knowledge. Whether aviation took a different course as a result of Watson's work can be debated. The achievement most deservedly attributed to him is the invention of the rocking wing system of lateral control (sometimes referred to as the Parasol wing) which must have enlightened and informed the field of aviation. Although the aileron, which now prevails, was on the cusp of being another practical solution to the problem at the time, wing-warping certainly waned after Watson's invention. How much that was influenced by Watson's work is a matter of opinion.

Along with the rocking wing system, Watson had developed a 3-axis control operated by a single 'stick' with which to control climb/descend, turn and bank.

Was he the first to fly? However one cares to define 'fly', the answer is probably "No". If any convincing 'hop' (i.e. a straight line trajectory higher off the ground and further over it than could be attributed to the energy of the launch) is taken as the definition, then he certainly was not the first. If the definition used is any flight equal to, or in excess of, what the Wright Brothers achieved in December 1903, then it seems quite likely that the answer is still "No". He may have been ahead in the process of developing a better means of lateral control than the Wrights, but it seems very likely that he did not have a powerful enough power unit installed by that date. A familiar story, for if Percy Pilcher had had such a power unit before his untimely death in a flying accident, it is likely he would have been airborne with powered, manned, controlled flight before either Watson or the Wright Brothers.

Gibbs-Smith and others have set out to show that Watson could not possibly have progressed any of his designs without prior knowledge of what the Wright brothers had achieved.

After all the correspondence with James Y. Watson, and after J.Y. Watson's death, Gibbs-Smith wrote in the definitive volume of 1960 [59], "and he only succeeded in achieving an ingenious but impractical type of aircraft that has borne no fruit in flying history".

Perhaps he was overlooking the single control column?

Although Watson and the Wright Brothers may have been in touch with each other, it is very evident that Watson had a brilliantly creative mind, which attached to the scientific knowledge he had of Fluid Mechanics and his study of nature, enabled him to design a very practical aircraft from about 1902 which was only awaiting a suitable power source to enable him to launch it into the air.

His 'rocker' wing which enabled him to turn safely without the dreaded 'aileron drag' was a creation of sheer genius, which has been totally unrecognized. He had established the principle on which the hang-glider and weight-shift microlight aircraft turn-and-bank function is based.

As far as the Watson story is concerned, exact dates and aircraft performances are unlikely ever to be forthcoming and any records that he might have kept have not survived. J.Y. was married in 1921 and moved into "Balgowan", the family seat shortly thereafter. Ironically, it was then that he himself cleared out and destroyed most of the material recorded at the time. Apparently his mother was very displeased about it. What a pity that J.Y. had no prescience of his endeavours thirty-odd years later! The controversial issues may never be completely resolved.

But this is also a story of James Yeaman Watson and his struggles to have his brother recognised as a pioneer aviator. But for J.Y.'s efforts much of the information that is available would have been lost. What he collected and the way he used it to argue with established authorities are what has preserved it. But more than this, it has painted in some of the background of the sort of difficulties and rewards that many of the pioneers must have experienced: the advantages and disadvantages of public interest, inertia in various hierarchies, misreporting in the press, poor provision of resources and facilities and the tremendous resourcefulness and determination required to overcome fear and frequent disappointments in order to attain the objectives.

J.Y. summarised it thus:

If it were not for the fact that although my memory of my late brother's efforts to fly in 1903 is weak so far as a definite date is concerned, I do believe from the evidence contained in this album that 1903 was the year that both of us, after many attempts, were airborne for quite a number of times for a few minutes.

I have a clear recollection of the blacksmith's anvil plus two 56 lbs weights being hoisted by block and tackle, these weights being then attached to a pulley which hung high up on a tree.

Preston built three aeroplanes each costing his father about £1,000.

I had little interest or was any help in the building or designing of these aeroplanes, but being of lighter build by some 28lbs I think my flights in the 1903 machine were more sustained for a slightly longer period than Preston's.

Preston's three aeroplanes were all of the tractor type (propeller in front), and that fact alone I consider gave Preston and his companions a better chance of successfully handling these very early aeroplane than other builders, including the Wright Brothers, who worked on the pusher type.

Preston was married to Beatrice Philip in December 1906. Preston was twenty-six years old on his marriage day, his wife being of twenty-four years of age.

The first flight of his second plane took place at Errol.

I have a very clear recollection being asked by my brother to try out this aeroplane. Preston told me he had promised his wife not to fly it.

He further said, and it is a clear recollection, "Take care what you do, this machine will go right up". I think I smiled, got seated in the plane, the engine was started and I took off. To my astonishment with the stick full forward the plane continued to climb, and for the first time I saw a neighbouring wood of trees from a height I would guess at 200 ft.

My effort then was to make a landing by cutting the engine revs. The landing was heavy but the under carriage stood the shock. The plane was damaged having turned on its nose, and the propeller broken. While I was disentangling myself from the damaged aeroplane, it was evident he was delighted with the performance of his second plane.

This plane was flown successfully by quite a few 'would be' pilots, these flights having been made after adjustments had been made to the tail of the aeroplane, which on its first flight was found to be either at a wrong angle or too heavy.

Preston's third plane I personally never saw, being in America at the time Preston was competing in France with his third machine. In conversation with Preston's widow she has told me, she has clear recollections of Preston shooting gulls and the endless care he took in the examination of their wings, tails, feet and heads. It is evident that the building of his three aeroplanes was to a considerable extent modelled on the shape of these gulls.

My own recollection of Preston's study of gulls, while they were flying is as follows:-

While a number of his rugby friends were training for important matches, Preston sat on the wall at the Esplanade where the Tay Bridge enters the Tay River from the Dundee side; when he was asked what he was doing sitting while we were training, the reply was, "Someday we will fly like those gulls". His rugby friends thought at that time, that Preston was a bit 'queer'. This was about 1899 to 1900.

The contents of this album are (sic) the result of an effort to deal justly with a determined attempt by Preston Watson to use the air as a means of travel. I have felt it my duty to put together a detailed account of his success in building a machine which can be described as a powered airborne craft, in the days when the very idea was scoffed at.

Having completed this work I feel I have fulfilled my duty as a brother.

This was written in July 1955, after quite a lot of information gathering but before the real 'battle' of the claim had fully developed. But it could serve well as a statement of his final intuitive feelings on the whole issue because although he had seen fit to alter some of his original datings in the light of his arguments with Gibbs-Smith, he obviously still believed after 1957 in the originality and merit of what his brother had done.

J.Y. spent a lot of his later life in **his** determined efforts. He took the reversals fairly and squarely, tried to maintain intellectual honesty and was prepared to alter his opinions in the face of hard evidence; and not to waver if he thought the case against them was not strong enough. It was something of a 'David and Goliath' struggle in which he acquitted himself well.

It must have seemed to him that he had lost the contest but he could console himself with one thing. Negative publicity is still publicity and the publication of his activities at some length in the core literature on the subject, even if it was to refute the claim, meant that Preston Watson's name is there. It seems that the desire to be recognised to have been 'FIRST' remains the subject of much determined competition at an international level, even if the inter-individual rivalry fades with the passage of time and the demise of the principal players.

James Yeaman Watson died on 17 October 1957, just three days after Kerr B. Sturrock. His obituary [94] in the press read thus:

AIR PIONEER STILL FLEW AT 78

Mr James Y. Watson, seventy-eight year-old Dundee-born aviation pioneer and one of the first to fly, died suddenly yesterday at his Blairgowrie home.

Only a few days ago he had flown an aircraft from Scone aerodrome – home of Strathtay Aero Club, of which he was the oldest pilot member – on one of his frequent flights.

Mr Watson had been gardening at his home, "Heatherways", Rosemount shortly before his death. He died soon after he went back to the house feeling unwell.

James Y. Watson assisted his brother Preston, killed in a 1915 air crash, in achieving powered flight in a heavier-than-air machine in the early years of the century.

For the past few years he spent his time amassing data in an effort to establish the date of his brother's first flight.

Many think the first flight was in 1903 – several months before the Wright Brothers became airborne – but J.Y.'s tenacious search was inconclusive.

A descendant of one-time Provost Yeaman of Dundee, Mr Watson was the son of the late Thos. Watson, partner in the firm Watson & Philip, merchants. Educated at Dundee High School, he became a director of the firm.

FLEW REGULARLY

Mr Watson served in the First World War in the RFC and retained the interest in aviation he had first developed in 1903 for the rest of his life. He flew regularly from Scone though for the last few years his doctor insisted that he be accompanied by a second pilot on his flights.

Mr Tom Blyth, ex RAF instructor at the club, said last night 'I'm shocked to hear of his death. He was in great fettle last week. He always said that it was flying that kept him alive. He was a man I had the deepest regard for. The fellows all used to say they'd be happy if they were half the man he was at his age.'

A newsletter was sent to club members at the weekend to announce this year's winner of the J.Y. Watson Cup. This trophy goes to the pilot with the highest aggregate points for summer flights.

James Y. Watson, like his brother Preston was an athlete in his youth and both were in the Newport team that won the North of Scotland Challenge Cup in 1889.

During his brief lifetime, Preston built three machines and James, two stones lighter than Preston, made longer hops in the early aircraft. In a recent conversation, James told me of these flights.

"I have a clear recollection of being asked by my brother to try out a plane. Preston told me he had promised his fiancée not to fly it. He further said, and it is a clear recollection 'Take care what you do. This machine will go right up.'

I think I smiled, got seated in the plane, the engine was started and I took off. To my astonishment, with the stick full forward the plane continued to climb and for the first time I saw a neighbouring wood from a height I would guess at 200ft.

My effort was to make a landing by cutting the engine revs. The landing was heavy but the undercarriage stood the shock. The plane was damaged, having turned on its nose, and the propeller broken. I only hurt my leg.

The propellers for these planes were made by Mr Kerr B. Sturrock, Dundee joiner and shop-fitter, who died last week.

CATAPULTED

The planes were catapulted into the air at full throttle by a system of weights and pulleys. "It was very sudden", James would recall, "you had to hold on tight or be left behind."

In December 1953, a few days before the celebrations of the 50th anniversary of flight Mr Watson went to London to present his evidence. But the evidence was not recognized.

A few months ago, Mr Watson presented a handsome album with the history of his brother's career to Dundee art Galleries and Museums.

If this publication achieves its aim of securing a more appropriate place in aviation pioneering for Preston Albert Watson, then the authors would regard it as a joint venture between themselves and the WATSON BROTHERS.

References and Bibliography

[1] TAYLOR, J.W.R., (ed.)(1984) *The Story of Flight*, Chancellor Press, London.

[2] CAYLEY, Sir GEORGE, (1809*) On Aerial Navigation I,* Nicholson's Journal, no.24, pp. 164–174.
(1810), *On Aerial Navigation II,* ibid no. 25, pp. 81–87.
(1810), *On Aerial Navigation III,* ibid 25, pp. 161–169.
See also
(1962) GIBBS-SMITH, C.H., Sir George Cayley's *Aeronautics* 1796–1855, HMSO, London.

[3] GIBBS-SMITH, C.H., (1960) *The Aeroplane. An Historical Survey* HMSO, London.
(1970) *Aviation. An Historical Survey from its Origins to the end of World War II,* HMSO, London.
(1974) *The Rebirth of European Aviation 1902–1908*, HMSO, London.

[4] CHANUTE, O., (1894*) Progress in Flying Machines,* Lorenz & Herweg, Long Beach, USA (1976 reprint).

[5] TOBIN, J., (2003*) First to Fly,* John Murray, London.

[6] HISTORICAL FLIGHT RESEARCH COMMITTEE GUSTAV WEISSKOPF www.weisskopf.de/index.html

[7] GIBBS-SMITH, C.H., (1960) *The Aeroplane: An Historical Survey of its Origins And Development,* pp. 207–8, HMSO, London.

[8] BLAIR CASTLE ARCHIVES, NRAS 980, bundle127, Blair Atholl, Perthshire.

[9] JENKINS, G., (1999), *'Colonel' Cody and the Flying Cathedral,* Simon & Schuster UK Ltd., London.

[10] ALLEN, J., (2002) *Wings Over Scotland,* pp. 12–18, Tervor Limited, Scotland.

[11] WEBSTER, J., (1994) *The Flying Scots,* p. 9, The Glasgow Royal Concert Hall.

[12] MUNSON, K., (1968) *Bombers 1914–19,* p. 101, Blandford Press Ltd., London.

[13] WYKEHAM, P., (1962) *Santos-Dumont: A Study in Obsession,* p. 47, et seq., Putnam, London.

[14] SANTOS-DUMONT, A., (1904*) My Airships,* The Century Co.

[15] WYKEHAM, P., (1962) *Santos-Dumont: A Study in Obsession,* p. 69, Putnam, London.

[16] HOFFMAN, P., (2003) *Wings of Madness; Alberto Santos-Dumont and the Invention of Flight,* p. 254, Fourth Estate, London and New York.

[17] SCHULMAN, S., (2002) *Unlocking the Sky,* p. 41, et seq., Harper-Collins, inc., New York.

[18] WILSON, G., (1966) *The Making of a Lord Provost,* p. 51, David Winter & Son, Dundee.

[19] JENKINS, G., (1999) *'Colonel' Cody and the Flying Cathedral,* p. 52–56, Simon & Schuster UK Ltd., London

[20] Ibid., pp. 114–119

[21] CUNNINGHAM, T.F., (2007) *Your Fathers the Ghosts: Buffalo Bill's Wild West In Scotland,* Black & White Publishing Co., Edinburgh.

[22] DOW, B., (1996) *The Historical Journey of Watson & Philip 1873–1995,* Private publication by Watson & Philip.

[23] DUNDEE UNIVERSITY ARCHIVES

[24] ALLAN, J., (2003) *The First Flying Scotsman,* Scots Magazine, December 2003, pp. 624–631.

[25] BARNES, D.J., (2010) *Biographical Roll of Honour* as yet unpublished; personal communication.

[26] STURTIVANT, R., PAGE, G., (1992) *Royal Naval Aircraft and Units 1911–1919,* Air-Britain, Tonbridge, Kent.

[27] NATIONAL ARCHIVES, Ruskin Avenue, Kew, London.

[28] SMITH, R., (2002*) British Built Aircraft – Greater London,* p. 216, Tempus Publishing, Stroud, Gloucester.

[29] MUNSON, K., (1968) I. pp. 52, 130–132, Blandford Press, London.

[30] *AMERICAN JEWISH CHRONICLE EVENTS, 5697* (1918), 1 June 1918–31 May 1919, p. 217.

[31] GIBBS, W.R., (1954) *Aero-mechanical Analysis Lecture to Dundee Technical College.*

[32] *COURIER & ADVERTISER*, 26 February 1966, p. 4.

[33] OBITUARY, *Courier & Advertiser*, (1957) 15 July 1957, p. 11.

[34] SCHULMAN, S., (2002*) Unlocking the Sky*, p. 231, Harper-Collins, New York.

[35] ESNAULT-PELTERIE, R., *L'Aerophile*, January 1905.

[36] SCHULMAN, S., (2002) *Unlocking the Sky*, pp. 133–4, Harper-Collins Inc., New York.

[37] SMITH, A.S., co-author: personal communication from unpublished article.

[38] RILEY, J.F. (1957) *The First Flying Scot Meccano Magazine, vol. XLII, no. 6*, June 1957, pp. 284–5.

[39] LESLIE, J.D., *Daily Record*, 26 February 1949.

[40] *BLAIRGOWRIE ADVERTISER*, 29 January 1954.

[41] BOYD, M.B., (circa 1910) *The Conquest of the Air*, Dutheil Chalmers and Boyd Ltd, p. 36.

[42] Ibid., p. 7.

[43] WYKEHAM, P., (1962) *Santos-Dumont: A Study in Obsession*, pp. 268–269, Putnam, London.

[44] GOURLAY, J., *Dundee Courier & Advertiser*, 17 November 1955.

[45] *DUNDEE ADVERTISER*, 11 October 1909.

[46] DEMAUS, A.B., TARRING, J.C., (1989) *The Humber Story 1868–1932*, p. 81 et seq., Alan Sutton Publishing, Gloucester.

[47] *THE AERO* magazine, September 1909.

[48] ANZANI COMPANY HISTORY.

[49] WATSON, P.A., (1914) *The Watson Rocking Wing Aeroplane Flight*, 15 May, pp. 510–12, (see Appendix ii).

[50] OBITUARY, *Dundee Year Book*, 1915, pp. 84.

[51] LESLIE, J.D., *The Genius of Preston Watson Scots Magazine*, October 1953, pp. 8–13.

[52] CASELY, G., (2003) *A Hundred Years of Flight Leopard Magazine*, December 2003, pp. 25–27.

[53] ALLEN, J., (2002) *Wings over Scotland – a History of the Scottish Aero Club*, p. 12, et seq., Tervor Ltd., Scotland.

[54] STEWART, O., Tatler & Bystander, 11 November 1953.

[55] PROFUMO, J., Hansard revealed no recording of this, although many of the items relating to John Profumo obviously concerned civil aviation, including an answer concerning Errol aerodrome. It is likely that the mention of Watson to which J.Y. refers would have been in a speech made outwith Parliament. Wherever it was made, it was reported in the *Manchester Guardian* as taking place in November 1953. (See ensuing reference)

[56] *MANCHESTER GUARDIAN*, 15 December 1953.

[57] GIBBS-SMITH, C.H., *Aeronautics,* February 1954.

[58] STEWART, O., *Tatler & Bystander*, 6 January 1954, p. 28.

[59] GIBBS-SMITH, C.H., (1960) *The Aeroplane – An Historical Survey,* pp. 208–213, HMSO, London.

[60] GIBBS-SMITH, C.H., (1974) *Aviation,* p. 71, HMSO, London.

[61] MUNSON, K., (1969) *Pioneer Aircraft 1903–1914,* p. 129, Blandford Press, London.

[62] TOBIN, J., (2003) *First to Fly,* p. 228, John Murray, London.

[63] PEEBLES, P., (1998) *Model Aeroplane News,* June 1998.

[64] *READERS DIGEST, The Future of Flight?,* January 2012, p. 74, et seq.

[65] SCHULMAN, S., (2002*) Unlocking the Sky,* p. 215, Harper-Collins Inc.

[66] *DUNDEE DIRECTORY,* 1910–11, p. 131.

[67] *COURIER & ADVERTISER*, The Craigie Column, 15 January 1992.

[68] *PEOPLES' JOURNAL*, 7 May 1960.

[69] *COURIER & ADVERTISER*, 28 February 1960.

[70] DESOUTTER, D.M.,(1954) *All About Aircraft,* p. 229, Faber & Faber, Ltd., London.

[71] ALMOND, P., (1997*) Aviation* Konemann, Koln.

[72] ADAMS, L., *Daily Record*, 4 September 2003.

[73] *COURIER & ADVERTISER*, 11 December 2003, et seq., (correspondence).

[74] ADAMSON, P., LAMONT-BROWN, R., (1981) *Victorian & Edwardian Dundee & Broughty Ferry,* pp. 66–7, Alvie Publications, St. Andrews, Fife.

[75] *COURIER & ADVERTISER*, The Craigie Column, 9, 15 & 21 February and 5 March 1992.

[76] PETERKIN, T., *Scotland on Sunday*, 28 March 1999, p. 7.

[77] WATSON, N., *Scotland on Sunday*, 12 April 1999, p. 10.

[78] LESLIE, J.D., *Daily Record*, 26 February 1949.

[79] Ibid., 5 March 1949.

[80] *AEROPLANE*, 17 July 1915.

[81] *FLIGHT,* 3 July 1914.

[82] *L'AEROPHILE, Le Concours de la Securite*, 15 June 1914, p. 275.

[83] ESPITALLIER, G., *La Securite en Aeroplane, La Revue Aerienne*, 23 July 1914, p. 392.

[84] *L'AEROPHILE*, July 1914.

[85] MUNSON, K., (1969) *Pioneer Aircraft 1903–1914,* p. 112, Blandford Press, London.

[86] INQUEST REPORT, 1 July 1915 on Sub Lt Preston Watson, Lewes Castle Archives, East Sussex.

[87] *DUNDEE ADVERTISER,* 1 JULY 1915, p. 6.

[88] *DUNDEE YEAR BOOK* (1915), p. 14 .

[89] *THE PEOPLES' JOURNAL*, 3 July 1915, p. 9.

[90] *FLIGHT*, July 1915.

[91] *SUSSEX EXPRESS*, 2 & 9 July 1915, p. 12

[92] *SUSSEX DAILY NEWS*, 2 July 1915.

[93] *DUNDEE ADVERTISER*, 5 July 1915.

[94] OBITUARY, *Courier & Advertiser*, 17 November 1957, p. 6.

APPENDIX I

1909 patent and 1907 patent

Nº 47   A.D. 1909

Date of Application, 1st Jan., 1909

Complete Specification Left, 24th July, 1909—Accepted, 16th Dec., 1909

PROVISIONAL SPECIFICATION.

Improvements in Flying Machines.

I, PRESTON ALBERT WATSON, of The Retreat, 370, Perth Road, Dundee, in the County of Forfar, Scotland, Merchant, do hereby declare the nature of this invention to be as follows:—

5   This invention relates to flying machines, the objects being to provide means for giving lateral stability and for steering, such means being positively controlled by the operator.

In carrying out my invention I use a main aeroplane whose position is fixed relative to the propelling mechanism and the ballast, this plane being of the usual type.

10   Above this plane and carried by the frame of the machine is one or more lateral stability aeroplanes, each plane being capable of rocking about a fore and aft central axis.

If one stability plane is used the pivot is above the fore and aft central axis of the main plane and depending from the stability plane and at right angles
15 to it is a lever fixed to such movable plane. This lever is fixed in so far that it can tilt the plane to one side or the other, but it is free to vibrate in planes at right angles for a purpose now to be described.

The rudder (or rudders) which gives fore and aft stability consists of one or more planes, normally in a horizontal position and fixed to the frame so
20 that they can vibrate about a transverse axis. The controlling lever actuates the rudder by means of a pull and push rod or by two flexible wires on opposite sides of the pivot. The rudder is preferably at the rear, but another may be at the front. The ballast is preferably on a level with the lower or main aeroplane.

Dated the 30th December, 1908.

25

GEO. C. DOUGLAS & Co.,
Chartered Patent Agents,
41, Reform Street, Dundee,
Agents for the Applicant.

COMPLETE SPECIFICATION.

30   ### Improvements in Flying Machines.

I, PRESTON ALBERT WATSON, of The Retreat, 370, Perth Road, Dundee, in the County of Forfar, Scotland, Merchant, do hereby declare the nature of this invention and in what manner the same is to be performed, to be particularly described and ascertained in and by the following statement:—

35   This invention relates to flying machines, the objects being to provide means for giving lateral stability and for steering, such means being positively controlled by the operator.

In order that my said invention and the manner of putting the same into

[*Price 8d.*]

practice may be properly understood,. I have hereunto appended an explanatory sheet of drawings in which the same reference numerals and letters are used to indicate corresponding parts in the figures shown.

Figure 1 is a side elevation of a flying machine constructed and arranged in accordance with my invention. 5

Figure 2 is a plan and

Figure 3 is a view looking in the direction of the arrow A (Figure 2).

Figure 4 is an enlarged side view of the lever and its appurtenances for rocking the stability aeroplane and for operating the rudder.

Figure 5 is a view looking in the direction of the arrow B (Figure 4). 10

Figure 6 is an enlarged side view of the rudder.

In carrying out my invention I use a main aeroplane 1 whose position is fixed relative to the propelling mechanism 2 and the ballast, this plane being of the usual type.

The rigid frame work of the machine is indicated by the letters *f* and the 15 tension wires by the letters *t*.

Above this plane 1 and carried by the frame of the machine is a rocking aeroplane 3 capable of rocking about a fore and aft central axle 4. The pivot 4 is above the fore and aft central axis of the main plane and depending from the rocking plane and at right angles to it is a lever 5 fixed rigidly to such 20 rocking plane 3. This lever is fixed in so far that it can tilt the plane 3 to one side or the other, but it is free to vibrate in planes at right angles for a purpose now to be described. As in all aeroplanes, the front edges of these aeroplanes 1 and 3 are higher than the trailing edges relative to the flight path, so that during forward motion the aeroplanes support the machine. 25

The rudder (or rudders) which gives fore and aft stability consists of one or more planes 6, normally in a horizontal position and carried by the frame so that they can vibrate about a transverse axle at 7. The controlling lever 5 actuates the rudder by means of a pull and push rod or by two flexible wires 8 on opposite sides of the pivot. The lever 5 is compound being able to rock the 30 plane 3 by means of the standards 9 which are rigid with the stability plane 3 such standards carrying the pivot 10 about which the lever 5 can vibrate in a fore and aft direction. The rudder is at the rear.

When in flight the machine is guided in the following manner :—

The action of the rocking aeroplane is as follows :— When the rocking 35 aeroplane 3 is tilted out of the horizontal about the axle 4 by moving the lever 5 to one side, the normal pressure of the rocking aeroplane 3 is inclined out of the vertical and gives rise to a horizontal component pulling its axle 4 to one side relative to the line of flight. The frame *f* and the fixed plane 1 are thus caused to rotate about the line of flight, that is to say the plane 1 becomes 40 tilted about the line of flight and out of the horizontal. The normal pressure of the fixed plane 1 is thus inclined out of the vertical and gives rise to a horizontal component pulling the frame *f* to one side of the line of flight. The tail planes 6 are usually horizontal in a fore and aft direction and therefore have no normal pressure and consequently do not give rise to a horizontal component 45 although tilted round the line of flight. The tail planes do not therefore pull to one side of the line of flight whereas the frame *f* is so pulled. The frame *f* being pulled to one side moves to that side that is the direction of flight is altered because the resultant of the side movement with the original forward movement is a new direction of flight. The tail being at the posterior end 50 brings the longitudinal axis of the machine into the new direction of flight in the same way as the feathers of an arrow keep its longitudinal axis in the line of flight. Therefore the machine turns to one side. It is thus shown that by moving the lever 5 to one side the frame *f* can be made to rotate about the line of flight and that the machine can be made to turn to one side, that is to say 55 lateral stability and steering right and left can be controlled by means of the lever 5.

Nᵒ 47.—A.D. 1909. 8

Should the aeronaut desire to ascend or descend when he is moving in a straight line the lever 5 is moved fore or aft thus causing the front edge of the rudder to be moved downwards or upwards. Again, if he desires to turn to the right or the left he moves the lever to the left side or to the right, thus causing
5 the plane 3 to incline relative to the main plane 1 and thus as described above turning the machine to the right or the left. As the direction of motion of the lever can be compounded from these two motions, the plane 3 and the rudder can be moved so as to cause the machine to move up or down while at the same time moving to the one side or the other, that is to say by simply moving the
10 hand which actuates the lever in any desired direction the machine can be steered in any desired direction and the trim altered. By having the rocking plane 3 considerably above the fixed plane 1 it can be of small size and easily tilted.

By such an arrangement the machine can be guided with the least possible
15 mental strain and the least possible work. No increased resistance on one side to forward motion is caused when preserving lateral stability as is the case when the wings are warped or supplementary side surfaces are brought into play. Therefore no vertical rudder is required. The wings may be built rigid and can be kept constantly at the most efficient angle for aeroplaneing. Further
20 there is no side slipping during turning, the "banking" being automatically correct.

It has been shown that by having a rocking surface above a fixed plane steering right and left and preservation of lateral stability is accomplished.

Having now particularly described and ascertained the nature of my said
25 invention and in what manner the same is to be performed I declare that what I claim is :—

(1) In aeroplanes the use of a rocking plane situated on a higher level than the main plane, for preserving lateral stability and for steering right and left and controlled by a lever which also operates the horizontal rudder as
30 described and illustrated on the drawings annexed.

(2) In aeroplanes the combination of a fixed main plane with an upper rocking plane as described and for the purposes set forth.

Dated this 23rd day of July, 1909.

35

GEO. C. DOUGLAS & Co.,
Chartered Patent Agents,
41, Reform Street, Dundee,
Agents for the Applicant.

Redhill: Printed for His Majesty's Stationery Office, by Love & Malcomson, Ltd.—1909.

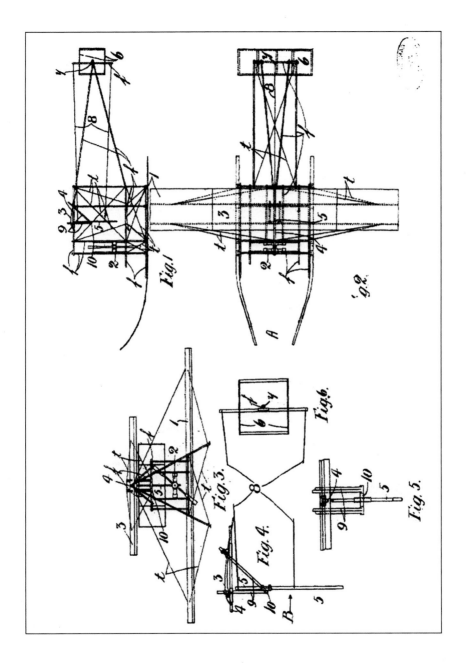

N° 23,553  A.D. 1907

Date of Application, 25th Oct., 1907

Complete Specification Left, 23rd May, 1908—Accepted, 8th Oct., 1908

PROVISIONAL SPECIFICATION.

Flying Machines.

I, PRESTON ALBERT WATSON, of The Retreat, 370 Perth Road, Dundee, in the County of Forfar, Scotland, Merchant, do hereby declare the nature of this invention to be as follows:—

This invention relates to flying machines the object being to raise and propel
5 a machine through the air without aerostats.

When birds fly the arms carrying the wing surfaces have a vibratory motion. One principle of this invention is that the arms carrying the wings are to rotate. The axis round which the arms carrying the wings rotate will be called the main axle. This main axle is in the same plane as the line of
10 direction of flight and is nearly parallel to it. In the case of horizontal flight the main axle is quite horizontal, but when the flyer mounts or descends the main axle although still in the plane of flight is inclined to the horizontal. In order to explain in which direction this inclination takes place and also for the further elucidation of the machine let me suppose that a man stands
15 in the flying machine facing the direction of flight, then a point passed by him outside of the machine on his right is said to be on the right hand side. If the flyer were flying in an oblique direction forwards and upwards and a man were standing on the right hand side on a level with the flyer and facing it as he flew past from left to right, the man standing on the right hand side
20 would see the main axle turned round slightly from the horizontal in the direction opposite to that of the hands of a clock. In the same circumstances if the flyer were flying in an oblique direction forwards and downwards the man standing on the right hand side would see the main axle turned round slightly from the horizontal in the direction of the hands of a clock. Although
25 the wings are carried bodily round an axis parallel to the direction of flight by the arms, a line of section of the wings at right angles to such direction is not rotated by the arms but remains horizontal relative to the machine. When the arms carrying the wings are driven upwards the planes of the wings feather round a horizontal axis at right angles to the direction of flight
30 to the air streams. When the arms bearing the wings are driven downwards the planes of the wings are kept from feathering round this axis by means of stops or bars.

Because the planes of the wings do not turn about an axis parallel to the direction of flight, in their rise and fall the air streams alternately strike
35 their upper and under sides respectively, and therefore these bars or stops while allowing wings to feather as they rise, prevent the wings feathering as they are driven downwards.

I shall now describe the apparatus for keeping horizontal the wings line of section at right angles to direction of flight, and for the feathering to the air
40 streams of the wings round a horizontal axis at right angles to the direction of flight.

Three arms of equal length are attached rigidly 120° apart to the main axle and in a plane at right angles to it. Three other arms of equal length are similarly attached to the main axle some distance along it. The plane con-
45 taining the first three arms and the plane containing the second three arms are

[Price 8d.]

Watson's Flying Machines.

therefore parallel. The arms in the one plane are opposite the arms in the
other plane, that is to say a plane containing the main axle and an arm
pointing upwards in one plane would contain an arm (pointing upwards) in
the other plane.

Three axles parallel to the main axle are carried loosely by the ends of the 5
arms further away from the centre of their rotation. These three axles are
called the secondary axles. When the main axle rotates these secondary axles
are carried round bodily by the arms but are kept from rotating themselves
by the following arrangement. The three secondary axles penetrate the three
arms which are in the plane at that end of the main axle towards the direc- 10
tion of flight and after passing through these arms are fitted with cranks.
The main axle is hollow and through it passes a rod, round which the main
axle revolves which is fitted with a crank in the centre of the three secondary
axle cranks, these four cranks being in the same plane. The rod passing
through the main axle, and the crank at the end of this rod remains 15
stationary. The four cranks while parallel are now connected by a link motion
so that the three secondary axles while having a motion of rotation about the
main axle do not rotate relatively to it or to each other. The edges of the
wings towards the direction of flight are called the anterior edges, and those
edges opposite them are called the posterior edges. 20

A cross bar is rigidly fixed to each of the secondary axles close to the three
arms which are adjacent to the cranks but on the opposite side of such arms.

These bars are rigidly fixed so as to be horizontal relative to the machine
and because the secondary axles do not rotate these bars remain horizontal,
relative to the machine. The anterior edges of the wings are fixed to these 25
bars, the plane of the wing being beneath the secondary axle. The wings
will therefore feather to the air streams when the secondary axles are rising.
The secondary axles themselves can be used as stops to prevent the wings
feathering when the secondary axles are descending. The stops of whatever
kind are placed so as to give a proper slant to the wings. This slant can be 30
varied by changing the position of the ballast. The rod terminating in a
crank which passes through the main axle forms a rigid support for it. The
hollow main axle is caused to rotate by means of a chain drive or by gear
wheels at the end further from the front.

The above apparatus is duplicated and one of each revolves on either side 35
far enough apart to prevent downward driven air striking the engine and both
somewhat above it. The two are exactly opposite each other relatively to
direction of flight, that is to say a line joining the centre of the crank on the
rod passing through the main axle on the right side to the centre of the crank
on the rod passing through the main axle on the left side, will be at right 40
angles to the direction of flight.

The lower part of the circles of rotation of the arms carrying the wings on
both sides, rotate away from the engine, they therefore rotate in opposite
directions. The wings are actuated by an engine driving two shafts rotating
in opposite directions. The engine is supported from the centres of rotation 45
by a triangular frame work and the wheels to run on the ground if such be
used by a rectangular frame work. Each wing may consist of a series of
smaller planes.

THE ADVANTAGES OF A MACHINE CONSTRUCTED IN ACCORDANCE WITH MY INVENTION.

The keeping of the wings line of section at right angles to direction of flight 50
horizontal.

The saving on this account of power by not having to turn the plane of the
wings about an axis parallel to the line of flight.

The saving of the power lost by a bird during the first and last portion of
their flap when the wings work against each other. 55

The natural gradual feather of the wings over the top part of the revolution of the arms carrying them,

N° 23,553.—A.D. 1907.   3

Watson's Flying Machines.

The ability to settle easily during flights the angle of the wings by simply putting ballast by trial so as to give horizontal or other flight.

The beat and propulsion when line of section of wings parallel to line of flight is almost horizontal gives the required greater lift than propulsion and
5 in that same position gives propulsion at a high gear with low speed of wings because air friction being the only obstacle to progression, while wings beat down a certain distance, the movement obtained forward is much greater.

The ability to allocate power to lift or propulsion by simply shifting ballast backward or forward power taken from the one being automatically given to
10 the other no loss of power being possible.

Air left after being struck by the wings is all going downwards or backwards, none upwards, as when using a screw for propulsion and none forward, as when using an aeroplane.

The new air obtainable by forward movement.
15 The new air obtainable in a zigzag downward motion.

The saving of power by the rotation of the wings instead of by vibration.

The small wing surface necessary compared to the aeroplane because the size of the wing is immaterial in this machine, smaller wings only requiring
a higher rate of revolution, no power being lost in the latter case.
20 The triangular form of support obtainable from centres of rotation to engine.

The two centres of support close together and the ballast almost immediately beneath, thus curing instability.

The saving of the power lost by driving a screw in the current of air caused by its own revolution, the propulsion in this case being obtained by striking
25 with slanted wings new air continually.

The small arc of feather necessary owing to the rapid rate of progression.

The three wings giving constant lift, thus preventing loss of power by the engine (which would be of very light moving parts) accelerating until all its power is absorbed in its own motion.
30 The hover possible by shifting ballast.

The hover possible by giving all the power to lift as against the power always required for propulsion of an aeroplane.

The ability to steer upwards, downwards and sideways, without a tail by shifting the ballast.
35 The ability to run on the ground.

The ability to aeroplane when wishing to do so or when compelled to do so as in case of an accident to the engine by stopping the wings, every one of the six acting as an aeroplane.

Dated this 24th day of October 1907.

40   GEO. C. DOUGLAS & Co:—
                                                Chartered Patent Agents,
                                                41 Reform Street, Dundee,
                                                Agents for the Applicant.

COMPLETE SPECIFICATION.

Flying Machines.

I, PRESTON ALBERT WATSON, of The Retreat, 370 Perth Road, Dundee, in the County of Forfar, Scotland, Merchant, do hereby declare the nature of this invention and in what manner the same is to be performed, to be particularly described and ascertained in and by the following statement:—

50   This invention relates to flying machines the object being to raise and propel a machine through the air without aerostats.

Watson's Flying Machines.

When birds fly the arms carrying the wing surfaces have a vibratory motion
while the means adopted in this invention consists in giving the wings, while
not themselves rotating, a motion of circular translation in a plane at right
angles to the direction of flight and causing them to feather to the air stream
while rising but, preventing their feathering while being driven downwards. 5
I thus provide a heavier than air machine, which will accomplish aerial flight,
which will have more stable equilibrium both transversely and longitudinally
and which will accomplish flight with greater economy of power and better
steering control, both transversely and longitudinally than heretofore.

The ordinary aeroplane with upward tilted anterior edge impelled forward 10
by a screw drives the air partially forwards and the normal pressure of the
aeroplane is partially backwards relative to the direction of flight. It is, in
fact, the method adopted by birds when they wish to stop or partially arrest
their forward motion. The air driven forwards by the aeroplane itself is the
cause of lost power, because this impelling of the air in a forward direction 1.
is quite unnecessary in flight. The back pressure caused by the forward
driven air also increases the "slip" of the screw. If instead of this aeroplane
motion, flight is accomplished by means of wings driven downwards successively
with their anterior edges tilted downwards so as to give an upward and forward
pressure, the air is driven downwards and backwards and inasmuch as upward 2
and forward pressure are both wanted in flight, there is in this latter method
no unnecessary power used. Further a great loss of power in aeroplaning
takes place at the screw by reason of "slip" and because of the air being
rotated and driven off radially in quite a useless direction. The loss is all
saved by birds in their flight, because the air is all driven in a direction which 2
gives useful pressure. The bird's method of flight consists of a succession of
glides, inasmuch as each beat of the wings is a glide. This method will be
referred to hereafter as successive gliding. A worse evil inherent in the
ordinary aeroplane than the waste of power is its instability. Stability in
flight may be preserved in the same way as the stability of an arrow in flight is 3
preserved. The bird's tail acts exactly as do the feathers of an arrow. In
preserving stability by this method the essentials are that the propelling force
act on the anterior end of the machine and that the surface which acts as a
tail and all other resistances to the air act on the posterior end of the machine.
If any force then tends to upset the equilibrium of the machine, the tail or 3
other resistance to the air is displaced from the line of flight and the air
stream acting on it tends to put them back into the line of flight. In the case
of an arrow, the propelling force is the momentum of its weight and the
centre of gravity in an arrow is always towards the anterior end. In the case
of bird or mechanical flight therefore, the propelling force should be made to 4
act on the anterior end and all resistance to the air, as far as possible, should
be at the posterior end. In an aeroplane this is not the case. The aeroplane
surface is tilted up relative to the direction of flight, offering resistance to the
air and as the aeroplane surface must be vertically above the weight, this
resistance is not at the posterior end of the machine; and the machine is very 4
unstable. In contra distinction to the action of the aeroplane is that of successive gliding in which case the normal pressure of the wings is forward
relative to the direction of flight and therefore the wings instead of offering
resistance to forward motion drag the machine forward and, with a tail
behind, the bird or machine is perfectly stable. Lateral stability in succes- 5
sive gliding is best preserved by shifting ballast sidewise, but it can also be
effected by rocking the wings transversely. Steering towards either side of
the direction of flight in successive gliding may be done by first shifting the
ballast aft and, then either shifting the ballast to the side to which it is
desired to turn, or rocking the wings towards that side. t

When it is desired to stop the motion of a successive gliding machine, the
posterior edges of the wings may be pulled down or the ballast may be shifted

back and the machine then acts as an aeroplane and thus the forward motion
is gradually arrested and a safe landing made. If, in successive gliding the
motor accidentally stops, a safe descent can be made by this aeroplane motion.
The machine is best steered for descent or ascent by shifting the ballast fore
5 or aft, but this can also be effected by tilting the tail upwards or downwards.

In order that my said invention and the manner of putting the same into
practice may be properly understood, I have hereunto appended five explanatory sheets of drawings in which the same reference numerals are used to
indicate corresponding parts in the figures shown.

10   Figure 1 (Sheet 1) is a view of a flying machine constructed in accordance
with my invention and looking in the direction of the arrow A (Figure 2).

Figure 2 (Sheet 1) is a plan and

Figure 3 (Sheet 5) is a side elevation of same.

Figures 4. 5. and 6 (Sheet 2) show the machine with an eccentric in place
15 of a crank on the main axle the motion being the same. These figures also
show how two arms 180° apart may be used in place of three arms 120° apart.

Figure 4 (Sheet 2) is a view of this arrangement looking in the direction of
the arrow B (Figure 5).

Figure 5 (Sheet 2) is a plan and
20   Figure 6 (Sheet 2) is a side elevation of same.

Figures 7. 8. 9. and 10 (Sheet 3) show the machine with a modified form of
framework the motion being the same. Here also the two arm form is shown.

Figure 7 (Sheet 3) is a view of this arrangement looking in the direction
of the arrow C (Figure 8) the dotted lines showing the position of the sails or
25 wings when the parts have made a quarter turn.

Figure 8 (Sheet 3) is a side elevation and

Figure 9 (Sheet 3) is a plan

Figure 10 (Sheet 3) is a plan of the machine when the wings have made the
quarter turn and are in the position shown by the dotted lines in Figure 7.
30   The figures on (Sheet 4) show enlarged details.

Figure 11 (Sheet 4) is an enlarged longitudinal sectional view of a portion
of a secondary axle showing how the swivelling of the sheet line is accomplished.

Figure 12 (Sheet 4) is a section at D. D (Figure 11) and
35   Figure 13 (Sheet 4) is a view looking in the direction of the arrow E
(Figure 11).

Figure 14 (Sheet 4) is a part longitudinal sectional view of the main axle
and hub showing the disposition of the two sheet lines and the mode of swivelling same.
40   Figure 15 (Sheet 4) is a section at F. (Figure 14).

Figure 16 (Sheet 4) is a side elevation of the controlling lever for manipu-
lating the wings and the tail or rudder of the machine illustrated by Figures 7
to 10

Figure 17 (Sheet 4) is a view looking in the direction of the arrow G.
45 (Figure 16).

Figure 18 (Sheet 5) is a modified form of frame for carrying the wing axles
and

Figure 19 (Sheet 5) is a modified form of upright for carrying the wings.

Figures 20 and 21 (Sheet 4) show a modification of the lever 19 adapted to
50 modifications of the machine illustrated by Figures 1 to 6.

The arrows H show the direction of flight. Parts of the machine towards
the front will be called the anterior and parts towards the rear the posterior.
Each set of wings may rotate in either direction so long as the other set rotate
in the opposing direction.
55   In carrying out my invention, with reference to the machine illustrated
by Figures 1. 2. and 3 (Sheets 1 & 5) I use a rigid framework 1 on which is
a platform 2 which carries the motor 3 and the aeronaut.

Watson's Flying Machines.

This framework is provided with wheels 4 for use when the machine commences to fly and a skiffing board 5 for preserving fore and aft stability when at rest on the ground. The motor drives the hollow hubs 6 by means of open and crossed chains 7 running on sprocket wheels 8 keyed on the hubs. Within each hub 6 and supporting them are cranked axles 9. These are referred to 5 as the main axles. They are carried in bearings 10 in the rigid frame work 1. These main axles do not rotate although they can be made to vibrate in their bearings. if desired. Each hub carries two sets of rigid arms 11 which have bearings 12 at their extremities and in these bearings are cranked axles 13 which are referred to as secondary axles. On each of these are fixed upright 10 rods 14 to which the wings 15 are pivoted at 16. Each set of three secondary cranked axles 13 are coupled to their main cranked axle 9 by connecting rods 17 and the two cranked axles 9 are coupled together by a rod 18 whose position can be regulated by the lever 19. By such an arrangement of parts the wings can be made to have a motion of circular translation about their 15 axles 9 and yet be kept in the horizontal position when the lever 19 is vertical, or they can be made to have a motion of circular translation in a transversely tilted position if the lever be moved to one side or the other.

The wings are allowed to feather naturally—that is the wings being free to vibrate about their anterior edges, their posterior edges being free the wings 20 take up a neutral position parallel to the air streams—while rising, but may be kept from feathering upwards above a certain point when the wings are driving downwards by means of a sheet line attached to the posterior edge of the upper wings at 21 and to the axle 13 while a connecting rod 47 keeps the under wing at the same angle, or the amount of feather upwards may be con- 25 trolled by sheet lines 20 which are attached to the posterior edges of the upper wings at 21, from thence passing round pulleys 22 and 23 on the secondary axles 13. The sheet line then passes inside the secondary axle 13 through a hole and is there attached to the swivel 24. This swivel and pulleys 22 and 23 are shown in Figure 11 (Sheet 4). The outer part of this swivel is prevented 30 from rotating by a feather 52 on the inner wall of the axle. The plunger 26 is the complementary part of the swivel the relative rotation taking place at 24. The plunger is attached to a sheet line at 45 which passes out at the end of the axle through a cap 48 attached rigidly to the arm 11 and carrying rigidly the tube 27 and the pulley wheel 28. The arm 11 is free to rotate on the 35 axle 13 which normally does not rotate. The sheet line pulls the plunger 26, which may be square in section, into the tube 27 while the spring 25 keeps the sheet line taut. The sheet line after passing round the pulley 28 passes down the arm 11, Figure 3 (Sheet 5) round pullies 29 and 30 and inside the hub 6 which is so constructed that except at the ends where it runs on the axle 9 there is 40 a space between the inner wall of the hub 6 and the outer wall of the axle 9, as shown in Figure 14 (Sheet 4). The sheet line is there attached to a hollow swivel 32 surrounding the axle 9, the outer part of the swivel being rotated by a feather 49 on the inner wall of the hub, the inner part of the swivel being kept stationary by a feather 50 on the axle. From the posterior end of the swivel a sheet- 45 line passes along between the hub and the axle and out through a hole in the bearing 34 which is rigidly attached to the axle 9 and round a pulley wheel 35 (Figure 2) along the framework round a pulley 46 and thence round a pulley wheel at the universal joint 51 down the lever 19 and is attached to an adjustable catch 36 on the lever. 50

A connecting rod 47 Figure 3 keeps the under wing at the same angle as the upper wing.

The tail 37 is jointed at 38 to rise or fall, its position being determined by two lines 39 which are attached to the lever 19 at 42 (Figure 20). This lever 19 is free to rock fore and aft on the pins 48 (Figures 20 and 21) and in doing 55 so manipulates the tail by means of the wires 39.

From the foregoing it will be seen that the lever 19 can regulate the tilt of the sails by moving the rod 18 to one side or the other the amount of "feather" on the sails can be adjusted by moving the catch 36 to which the sheet lines are connected and the tail can be operated by vibrating the lever in a fore and
5 aft direction. In vibrating fore and aft the lever moves in the slot 43 which forms part of the bar 18.

Although I have described how the machine can be manipulated by adjusting the transverse tilt of the wings the feathering of the wings and the manipulation of the tail yet the machine can also be steered without having special
10 means for feathering the wings or tilting the wings transversely and this can be done by means of shifting ballast, which, when moved behind the centre of support will throw the anterior edges of the wings up, converting it into an aeroplane, which in conjunction with a shift of ballast sidewise allow of the machine being guided upwards downwards or sideways.

15 The modification illustrated by Figures 4. 5. and 6 (Sheet 2) is similar to that already described except that two sets of sails at each side are used instead of three and as eccentrics 44 are employed instead of cranks the main axles 9 form part of the frame work and the eccentrics are partially rotated round the main axles 9 by the rod 18.

20 The sheet line in this case follows the course 21. 22. 23 (Figure 6) 28. 29. 35. 51 (Figure 5) thence down the lever and is attached as before to an adjustable catch on the lever 19, the swivelling arrangements inside the axles 13 and hubs 6 being the same.

The motions of the wings are exactly the same in the two types of machines
25 shown by Figures 4—6 and 7—10 on Sheets 2 and 3. but the two sets of wings in the machine shown by Figures 7—10 on Sheet 3 are situated so that the hubs 6 are in one line and the main axles which are hollow form one continuous tube 9 (Figure 8). On Sheet 3—Figures 7 to 10—the motor 3 drives the hollow hubs 6 by means of open and crossed chains 7 running on sprocket
30 wheels 8 keyed on to the hubs. These hubs 6 run on a main central continuous tube 9 which does not rotate. It is kept from rotating by means of the lever 19 which controls its position and consequently the position of the main cranks 9. By means of the connecting rods 17 the motions of the secondary crank axles 13 are made to synchronize with the motions of the main axle 9. Each hub 6
35 carries two sets of rigid arms 11 which have bearings 12 at their extremities which carry the secondary cranked axles 13. On each of these axles 13 are fixed upright rods 14 to which the anterior edges of the wings 15 are pivoted at 16 so that their posterior or free edges can vibrate, the anterior edges being always at right angles to the rods 14 while the latter by reason of the mech-
40 anism described above are always parallel to the arms of the cranks 9. By such an arrangement of parts the wings can be made to have a motion of circular translation round the main axle 9 while their anterior edges are kept in the horizontal when the lever 19—Figures 16 and 17—is vertical or they can be made to have a motion of circular translation with their anterior edges tilted
45 out of the horizontal if the lever be moved laterally to one side or the other. The wings feather as described with reference to the machine illustrated by Figures 1. 2. and 3.

In Figures 1 to 6 the two groups of wings—each moving round their main axles 9—are on either side of the centre of the machine; while in Figures 8.
50 to 10 one group of wings moving round its main axle 9 is. relative to the direction of flight. in front of the other group of wings moving round its main axle 9, this main axle being common to both.

The framework 1. Figure 8. has bearings at 10 in which the main axle may be partially vibrated by the lever 19 of which an enlarged drawing is
55 given in Figures 16 and 17 (Sheet 4). This lever 19 is attached to the main axle 9 so that when the aeronaut facing the anterior end of the machine moves the lever right and left, the main axle is partially vibrated and the wings are

Nº 23,553.—A.D. 1907.

Watson's Flying Machines.

thereby rocked transversely. The lever is free to rock fore and aft on the pins 48 (Figures 16 and 17) and in doing so manipulates the tail by means of the flexible wires 39 which run through the main axle and crank 9 emerging at 40 and 41 (Figure 8). The sheet lines in this case follow the course 21. 22, 23, 28, 29, 30, 48, (Figure 8) 36 (Figure 17), the swivelling arrangements 5 inside axles 13 and hubs 6 being the same as described with regard to the machines shown in Figures 1—3 or 4—6.

The arms 11 projecting from the hubs 6 may be strengthened by wire guys or a complete wire spoke wheel as shown in Figure 18 (Sheet 5) may take the place of the arms, the bearings 12 being attached to the circumference of the 10 wheel. The uprights 14 may take the form shown in Figure 19 so as to give greater strength.

Having now particularly described and ascertained the nature of my said invention, and in what manner the same is to be performed I declare that what I claim is:— 15

The mode of sustaining, propelling and steering flying machines by means of symetrical groups of feathering wings, each wing although not rotating having motion of circular translation, the one group circulating in the opposite direction to the other and the wings while always maintained in approximately a horizontal position being capable of a modified common and similar adjustment 20 from a central lever with respect to lateral tilt, the upward feather of the wings being limited by means of fixed sheet lines, or, if desired by sheet lines controlled at the central lever which lever also regulates the position of the tail as described and illustrated on the drawings annexed.

Dated this 22nd day of May 1908. 25

GEO. C. DOUGLAS & Co:—
Chartered Patent Agents,
41 Reform Street, Dundee,
Agents for the Applicant.

Redhill: Printed for His Majesty's Stationery Office, by Love & Malcomson, Ltd.—1908.

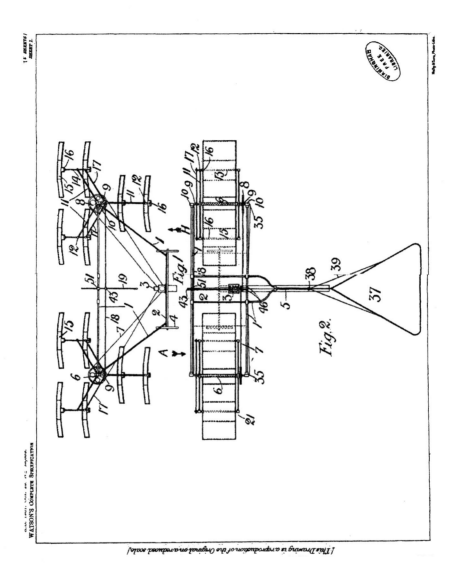

APPENDIX II

Flight magazine article of May 1915

MAY 15, 1914.

THE WATSON ROCKING-WING AEROPLANE.

In connection with the competition organised by "L'Union pour La Sécurité en Aeroplanes," Mr. P. A. Watson, of Dundee, who has been experimenting with rocking-wing aeroplanes for a number of years, is at present demonstrating his latest type machine (No. 3) at Buc. We have asked Mr. Watson for a description of his machine, and he has sent us a copy of the explanation

carries at the rear a monoplane elevator and a small fixed vertical fin. It will be observed that no movable vertical rudder is fitted. Mounted on a very strong *cabane* formed by two pairs of inverted V struts, is the balancing plane, by means of which lateral equilibrium is maintained, and the action of which is explained by Mr. Watson in the following extract :—

Side view of Watson No. 3.

which, in accordance with the rules of the competition, he has given l'Union. The following extract from Mr. Watson's explanation of the principles of his machine will, we think, make it clear in what respect his aeroplane and the manner in which it is controlled differ from ordinary types.

—The machine (No. 3), as will be seen from the accompanying photographs, is of the tractor type, and is driven by a 40 h.p. Anzani engine, mounted in front of the wicker-work *nacelle*. The chassis, which is of the four-wheeled type, is very robust and provides a comparatively wide track. An outrigger formed by two tail booms

"The method of preserving lateral equilibrium invented by the Wright brothers has been slavishly followed, but this has probably been due to the fact that these gentlemen were the first to fly in a practical way. This does not prove, however, that they have not misled everybody as to the best means of preserving lateral equilibrium. It must be remembered that they began their experiments at Kitty-hawk with the fixed intention of preserving lateral equilibrium by warping the wings, and when this means alone was found insufficient they never considered the possibility of using other means than warping, but looked for an addition to their warping wing machine and devised the vertically pivoted tail.

"The Wright Brothers have stated, and it is well known, that if the angle of incidence of the lower wing is increased, its resistance is also increased, so that the fore and aft axis of the machine turns about its vertical axis, away from the line of flight, and the lower

Front view of Watson No. 3.

510

PRESTON WATSON

wing loses its velocity, unless this is prevented by the movement of a vertically pivoted tail. In the absence of a vertically pivoted tail the loss of velocity of the wing whose angle of incidence is increased, causes it to lose its support, and it descends while the other rises. Thus the vertically pivoted tail is proved to be necessary if lateral equilibrium is to be preserved by the warping of the wings. The absence in a bird of the vertically pivoted tail proves that warping of the wings is not the method employed by a bird to preserve lateral equilibrium.

" The method by which a bird preserves its lateral equilibrium, and steers to right or left, is a beautiful method of flight. If a soaring bird is making a straight flight with its wings transversely in the horizontal and it wishes to steer to one side, it rocks its wings about a fore and aft axis by pulling one wing down and allowing the other to rise. It is able to do this because the centre of gravity of a bird is below its centre of support, and a pendulum thus exists. The force which this pendulum exerts if shifted from the natural position to which it hangs, provides a fulcrum in mid-air from which the wings can be rocked. Since the mass of the pendulum is considerable its inertia also helps in providing a fulcrum. When the wings are rocked out of the horizontal their lift has a component force pulling to one side of the line of flight. Now, when a body which is moving in a straight line is acted on by a constant force at right angles thereto, the body describes a circle. Therefore when the bird has rocked its wings it describes a circle.

" It must be remembered that this beautifully balanced flight takes place without any attention on the part of the bird except that it controls the elevation (according to whether it wishes to ascend or descend or fly horizontally during the turn) and rocks its wings to the extent which it considers necessary to make a circle of the desired diameter. In fact the bird possesses a method of flight which takes care of itself and is controlled by two movements. The difficulty which caused Wilbur and Orville Wright to abandon this beautiful method of flight adopted by the bird was that considerable power is required to rock the whole surface of the wings in the manner of the bird.

" A soaring bird has sufficient power to rock the whole wing surface quickly, since it employs for this purpose its strong flying muscles, but the pilot of an aeroplane has not sufficient power for this purpose. Thus the Wright brothers abandoned the perfect method of flight of the bird in favour of warping wings and a vertically pivoted tail, because with this latter method the pilot can preserve lateral equilibrium without having to exert so much power. Less power is required to warp or to control the aileron than to rock the whole wing surface in the manner of a bird. The choice between one or the other of these methods of preserving lateral equilibrium is a choice between the perfect flight of the bird, which, however, has the disadvantage of requiring considerable power on the part of the pilot, and the method of preserving lateral equilibrium by controlling the angle of incidence of the wing tips, or the angle of incidence of the supplementary surfaces, a method which has the advantage of requiring small power on the part of the pilot, but which causes an excess of resistance on the surface which has the greater angle of incidence, and thus necessitates the vertically pivoted tail with its consequent disadvantages.

" It remains to examine whether it is not possible to invent a method of preserving lateral equilibrium, which requires small power on the part of the pilot, and which does not increase the resistance of one side of the machine and thus does not necessitate the use of a vertically pivoted movable tail.

" In the machine described, a supplementary aeroplane surface possessing a lifting effect is situated above the main aeroplane, and

Three-quarter front view of chassis and nacelle of the Watson No. 3, showing slots in main planes through which the pilot raises his legs when lying down inside the nacelle.

is attached to an upward extension of the frame of the machine. This supplementary surface can be rocked about a fore and aft axis with the exertion of small power on the part of the pilot, and when thus rocked it gives rise to a component side force similar to the side pull of the wings of a bird when they are rocked. This side pull is exerted on the upward extension of the frame of the machine, and thus controls the ' list ' of the frame of the machine in the same way as a bird controls its ' list.' The main aeroplane is rigid with the frame, so that the rocking of the upper plane controls the rock of the main plane. The torque about the fore and aft axis of the machine depends on the distance between the centre of gravity of the machine, and the upper part of the upward extension of the frame where the supplementary aeroplane surface is situated. This surface may be of small area, and may still exert a sufficient torque about the fore and aft axis of the machine, if the upward extension of the frame is sufficiently long. It must be remembered that the pressure on the upper rocking wing is always balanced about the axle on which it rocks, so that the rocking does not require great power on the part of the pilot. On the other hand when wings are warped, the wing with the

Watson No. 2 in flight at Errol, Perthshire, 1912.

Watson No. 2, 1912, showing balancing plane in operation.

C 2

168

FLIGHT

MAY 15, 1914.

greater angle of incidence has the greater pressure, and considerable power is required on the part of the pilot because of this unbalanced pressure. By placing the supplementary aeroplane surface above the main aeroplane, and in making it to rock about a fore and aft axis, power is obtained with which to rock the main aeroplane surface. This is obtained without introducing any other force than the one required, that is to say the equilibrium of the aeroplane is left unaltered in every respect except that the 'list' is controlled. For instance, the relation between the resistance to the movement of advance, between the upper and lower parts of the machine, is not affected when the upper wing is rocked. When the main plane is rocked out of the 'horizontal' by the action of the supplementary

Three-quarter rear view of Watson No. 3, now at Buc.

plane, the machine makes a circular flight in the same way as a bird makes a circular flight. Thus the pilot can control the 'list,' and can steer to right or left as he desires. The resistance on one side of the machine is not increased when the main plane is rocked by the upper plane, since the main plane and the upper plane have everywhere the same angle of incidence. Thus there is never a tendency of the fore and aft axis of the machine to turn about the vertical axis, away from the line of flight. Therefore no movable vertically pivoted tail is required. Since there is no warping of the wings, and since no vertically pivoted tail is required, movements to create an exact balance between the warp and the vertically pivoted tail are not required. Steering to right and left is caused solely by the rocking of the wings out of 'the horizontal.' Therefore the 'banking' is always just as much as the turn requires. If for any reason the machine acquires a 'list,' the pilot has only to rock the upper wing to correct

this list, and if he desires to circle to right or left, again he has only to make one movement, namely, to rock the upper wing ; the main plane is then rocked by the upper wing, and the circle is made without requiring any other attention on the part of the pilot, except that he controls the angle of incidence of both the upper and main plane, by means of the elevation rudder so as to make an ascent, a descent or a horizontal flight, taking care not to make a rock of the main plane too great for the power of the engine. If the turn becomes too quick, as may happen because of the tendency of the outer wing to rise and thus to increase the rock of the main plane, the pilot has only to rock the upper plane so that the upper part of the frame is pulled away from the centre of the circle. The main plane thus approaches more nearly to the horizontal and the circle increases in diameter.

"My machine is almost 'fool-proof,' for, as has been shown, only one single movement must be made to preserve lateral equilibrium and to steer. This consists of a transverse movement of the lever. The elevator is controlled by a fore and aft movement of the same lever. The pilot, by rocking the upper plane, gradually rocks the main plane until the circular flight is being made as quickly as he desires, and the turn then takes place without requiring any other attention on the part of the pilot. In fact, this apparatus possesses the advantages of the bird's flight and requires little power on the part of the pilot. The equilibrium is natural compared to the forced equilibrium of the warping wing machine.

"There is another difference between the machine described in the enclosed specification and the machine with warping wings, which is probably *the greatest advantage of all* possessed by the former. This is, that the warping of the wings and the movement of a vertically pivoted tail, because of their indirect action, that is to say because they depend on the speed of advance, preserve their power to restore the equilibrium, only if the aeroplane has sufficient speed to make the warping of the wings and the movement of the vertical tail effective. If the ordinary aeroplane has lost its speed of advance by being made to ascend too quickly, or if it begins to sideslip, no amount of warping of the wings and movement of the vertically pivoted tail will prevent a tendency to capsize laterally, and even in calm weather there is a tendency to capsize laterally if a single propeller is used and the engine is kept running, because of the torque of the engine acting on the machine.

"On the other hand, the machine described in the enclosed specification may be made to preserve its lateral equilibrium, even if it has lost its speed of advance, because of the positive action of the upper plane in preserving lateral equilibrium. Thus if the machine has a 'list' to one side and has lost its speed of advance, so that it begins to descend, the upper wing, when rocked, is caught by the current of air caused by the descent, and the machine is 'righted.' In this case, the upper wing acts almost like a safety parachute, and not only 'rights' the machine but stops the descent."

✸ ✸   ✸ ✸

Airship Sheds at Farnborough.

IN answer to a question put in the House of Commons last week, Mr. Churchill stated that the airship sheds at Farnborough were only intended to house small airships for training purposes. It was under consideration to remove these sheds to another site, and if this were done it might be found convenient to slightly alter and improve them at the same time.

Hollow Wood in Aeroplane Construction.

IN our issue dated May 2nd, we published a translation of an article that had appeared in *l'Aérophile*, in which the method of hollowing struts and spars on the Farman and other machines was described. This concluded with a comparison between the strength of a hollow and a solid strut, stating how much superior the former is to the latter in its resistance to fracture under bending stresses. It should be noted, however, that whereas in the original article the dimensions are given in millimetres, the modulus of the section has been calculated from centimetre measurements, and that although the conclusions arrived at are, in general, correct, the strut is subject, in practice, to an end loading, and the transverse bending load is of negligible amount.

Blériot Brookland School Starts.

ACTIVE work at the new school which has been organised by the Blériot Co., at Brooklands, will commence on Monday next.

Gordon-Bennett Eliminating Trials.

FROM the official notices of the Royal Aero Club on p. 514, it will be seen that the eliminating trials to select the team of three to represent Great Britain in the Gordon-Bennett Race, will be held on Salisbury Plain during the last week in August. Each entrant will be allowed to fly three machines, and the programme will include slow and high speed tests over a distance of 200 kiloms. made up of 20 circuits of a 10-kilom. course.

Aviation in New Zealand.

REVERTING to the article which appeared in our issue of May 2nd and under the above title, it should have stated that the photographs were sent by Mr. A. W. Schaef, who also sent along the details of his own flying and those relating to the smash of Scotland at Newtown Park, Wellington. Incidentally it may be pointed out that only five weeks elapsed between the occurrence of the mishap and the appearance of the particulars in FLIGHT. It is interesting to learn that Mr. Schaef is now on his way to England, where he intends to secure his pilot's certificate and obtain as much information as possible, first hand, regarding matters aeronautical. Mr. Schaef is sure of a warm welcome in England, especially from those who are acquainted with the work which he has accomplished on the Anzani-engined monoplane which he constructed himself.

Acknowledgements

Our grateful thanks are due to many for help and encouragement along the route of compiling this book. We would like to mention in particular the following:

Fiona Sinclair and the staff of the McManus Collection in Dundee; Murray Thomson and D.C. Thomson Publications; the archivists of Dundee University, Abertay University and the Blair Atholl estates; Christophe Goutard of the Musee d l'Air at Le Bourget; James Watson – Preston Watson's nephew; John Gourlay; Ronald Patterson; James Allen; Gordon Lochead: Marie Muszynski – librarian to Fife Hospitals; Charles Philip of Watson & Philip; Alec Coupar of Spanfoto, Dundee; David and Judy Nichol of Forgandenny.

We also acknowledge the patience and forbearance of our respective wives to whom this modest publication is dedicated.

AWB
AS
March 2014

Index